NOBLE OBSESSION

NOBLE OBSESSION

Charles Goodyear, Thomas Hancock, and the Race to Unlock the Greatest Industrial Secret of the Nineteenth Century

CHARLES SLACK

AN IMPRINT OF HYPERION
NEW YORK

ISBN: 0-7868-6789-2

Hyperion books are available for special promotions and premiums. For details contact Hyperion Special Markets, 77 West 66th Street, 11th floor, New York, New York 10023, or call 212-456-0100.

Book Design by Lorelle Graffeo

FIRST EDITION

10 9 8 7 6 5 4 3 2 1

For my parents, Carolyn and Warner Slack

CONTENTS

CONTENTS

NOBLE OBSESSION

PREFACE

Rubber is everywhere today. It is as common as the sole of your shoe, the washer in your kitchen pipe, the elastic band holding your daughter's ponytail in place. Rubber has paid a price for ubiquity—we hardly give it a thought.

When we *do* think of rubber, it is certainly not on the order of a miracle. And yet that's exactly how rubber was greeted when it started appearing in large quantities in port cities of Europe and the United States during the early part of the nineteenth century. In a world dominated by wood, iron, steel, canvas, and leather, this bewitching substance that seeped from tropical trees was unlike anything else. It was pliant, airtight, and waterproof. It could be molded or spread on cloth. It absorbed shocks, would not conduct electricity, and could stretch to many times its length, only to snap back into its original shape.

Europeans had known about rubber and marveled at its properties since the early sixteenth century, but they never knew quite what to do with it. Rubber patiently awaited its time of glory. That time came with the dawning of the industrial age. The nineteenth century, with its newer and ever larger machines, cried out for just such a material. Although the first practical use for rubber was in clothing, engineers quickly realized its irreplaceable value in preventing a rapidly industrializing world from dashing itself to bits. Rubber springs cushioned the rides of railroad cars and carriages; rubber seals and gaskets prevented leaks and explosions; rubber belts made huge gears mesh smoothly; rubber packing kept machines from rattling apart. Rubber became the great shock absorber of the industrial age.

Likewise, few of the industrial developments that defined twentieth-century life (and continue to define life in the twenty-first) could have happened without rubber. The automobile's dependence on rubber for tires, hoses, belts, and seals is so well known that it almost goes without saying. Tires still consume half of the world's supply of rubber, both natural and synthetic. But electrical power, telephones, and modern plumbing could not have been brought to millions of people without rubber to insulate and seal. Powered flight could not have taken off without rubber. To this day, virtually every machine that rolls off an assembly line (usually on a rubber conveyor belt) contains crucial components made of rubber. Think of a world without phones, lights, plumbing, airplanes, and the family minivan. Think of a world without soccer balls, footballs, golf balls, basketballs, and tennis balls. Think of a world without soft mattresses, springy carpets, diving suits, running shoes, condoms, and balloons. Think of a world without all of these things and you have a world without rubber.

Actually, that might have happened.

From the start, rubber had a near-fatal flaw, one so vexing and stubborn that early experimenters puzzled over it for years to no avail. But for the obsession of one man in the United States, and his competitor in England, the world might be a very different place.

One thing needs to be said up front: Charles Goodyear, the American in this story, didn't start the tire company. The Goodyear Tire & Rubber Company was started by people he never knew, in a city he never visited, thirty-eight years after his death. Given the understandable (and not entirely accidental) confusion on this point, it bears mentioning that among the hundreds of potential uses for rubber envisioned by Charles Goodyear, the automobile tire was one he never foresaw, for the simple reason that the automobile had yet to be invented.

1

THIS SHIP OF AIR

Thomas Hancock sat in his London office on a late summer's day in 1842, turning over a few darkened scraps of material in his hands. To an untrained eye, they might have appeared to be strips of leather or bark. In any case, they were unprepossessing to say the least, not the sort of objects a busy man would be likely to waste much thought on. But Hancock was stunned. He pulled the strips at each end, bent them, rubbed his fingers approvingly over the smooth surfaces. He turned to his friend William Brockedon, who had given him the samples.

"Where did you get these?"

Brockedon explained that an Englishman named Stephen Moulton had brought them from New York on behalf of an unnamed American inventor. The inventor sought fifty thousand pounds in exchange for the formula and exclusive rights to manufacture. Hancock nodded and looked back down at the scraps.

He was no longer a young man. At fifty-six, he had devoted nearly a quarter of a century to the rubber business. The years showed in the heavy pouches under his eyes and the receding hairline that he masked with white wisps brushed over from the side of his head. He had started out as a carriage maker but had encountered rubber sometime before

1819 and been transfixed. His early discoveries had earned him the unofficial title Father of the Rubber Industry.

Nobody knew better than Thomas Hancock the endless commercial possibilities of rubber, if only its great flaw could be eliminated. Despite its seductive qualities, rubber could not withstand serious fluctuations in temperature. In very hot weather it turned sticky to the touch. After a time, it melted and ran into rank, shapeless blobs of foul-smelling goo. Cold had the opposite effect: goods turned hard and brittle and cracked like china plates. For years Hancock, with his rare combination of creative genius and plain business sense, had managed to keep his own business afloat when most others in England and the United States had failed. He did so by sticking to a narrow range of products that did not have to withstand severe conditions. But there was no way that rubber could be anything more than an industrial sideshow unless someone found the answer. Like many others, Hancock had tried, and failed. Nearing the twilight of his career, he had begun to believe that the answer did not exist.

And now, after all these years, came conclusive evidence that the answer *did* exist. By God, it *could* be done! And the proof had quite literally fallen into his hands. The scraps were smoother than any rubber he had felt before, and tougher and stronger. A few simple experiments confirmed his initial observations. Exposed to heat, the scraps retained their shape and smoothness; exposed to cold, their pliability. Solvents, which dissolved raw rubber, had no effect on these pieces.

Hancock could not help but feel joy in seeing this code cracked at last; he could not help but be impressed by the handiwork of the unnamed inventor. And yet any joy was tempered by the realization that someone other than he was responsible. How was it that the American had taken out no patent on such a discovery? How could he be so foolish as to show samples around without protection? Hancock himself knew how to keep a secret. He gave his own discoveries deliberately misleading names, and swore a few privileged employees to silence. Any experiment that Hancock considered sensitive he conducted not in his factory, but in a locked laboratory in the attic of his London house.

Now Hancock faced a great choice, the sort a man might face once in his life, when the stakes are the highest and the pressure against doing the right thing the greatest. The choice was clear—Hancock could seek out this obscure American, openly laud his achievement, then attempt to forge a business partnership that might benefit them both. Or he could take advantage of the inventor's carelessness, use his own extensive knowledge of rubber and his methodical mind to unlock the secrets contained in these samples, and claim the invention and its rewards for himself and his company. Perhaps it was no choice at all. Hancock's decision has ever since been the subject of debate, defensiveness, angry charges, and denials stretching across the Atlantic and back again. Thomas Hancock stared at the samples in his hands and saw the future of the industrial world staring back at him. His laboratory beckoned.

❁ ❁ ❁

In America, Charles Goodyear, forty-two years old, sick, tired, poor, and exhausted, eagerly awaited word from England. For the past eight years he had devoted himself with monomaniacal zeal to his quest. He had endured poverty, ridicule, life-threatening illnesses, jail, and self-torment in the knowledge that his wife and children had suffered as much or more than he.

Goodyear *was* rubber. He spent his days stirring it, boiling it, kneading it, reeking of it. He had transformed himself into a walking laboratory for all of his experiments. He wore caps, shoes, and coats made of the stuff. Goodyear reasoned that wearing rubber about the body was one of the best tests of its strength. Body heat tested its tackiness, while the constant motion tested its resilience. Often, in need of a swath of rubber on which to try some new solvent or treatment, a length of his coat would do. As a result, his coat sported large black or gray smudges. His strange attire made him an object of much ridicule as he wandered with his family from one town to another. Goodyear cared little what others thought of his appearance. He bore the ridicule cheerfully. He loved to repeat a story, perhaps apocryphal, of what a neighbor was supposed to have said when asked how to recognize Goodyear: "If you meet a man who has on an India rubber cap, stock, coat, vest, and shoes, with an India rubber money purse, without a cent in it, that is he."

He had overcome false starts and dashed hopes too numerous to remember. Long after his wisest friends had given up on him, Charles Goodyear endured. Now, at last, he had the answer. For years he had feared that he would die before the world recognized this essential truth: Charles Goodyear was no lunatic, but a visionary with a great gift to bestow on mankind. He felt certain that vindication was at hand, that he had won. He did not know that the race had just begun.

Charles Goodyear lived most of his life under a cloud of financial insolvency and desperation, due both to outside forces and to his own carelessness with money. He shamelessly sponged off of friends and family, pawned his loved ones' most cherished belongings, and languished in one debtors' prison after another. This chronic penury would have horrified Goodyear's earliest American ancestor, a thrifty, prosperous soul who helped found the New Haven colony and established the Goodyear name in the New World with prospects as bright as any and brighter than most.

In early 1637, a London merchant named Stephen Goodyear boarded the ship *Hector*, which, with a sister ship, carried 250 adventurers, merchants, and religious seekers to the New World. They were a particularly headstrong and ambitious bunch, even among the grand Utopian dreamers of the day. These were no humble sufferers for the Lord, seeking some barren rock on which to prostrate themselves while awaiting Judgment Day. They saw themselves as fathers of a vigorous new society, built on the twin pillars of righteousness and trade.

Finding Boston too confining and established, the party set its sights on the virgin expanse beyond. Sailing on down the coast, they came to a broad, welcoming harbor the Indians called Quinnipiac. They paid local Indians twelve coats, twelve spoons, twelve hatchets, twelve hoes, and twenty-four knives for the harbor and the land surrounding it. They renamed the spot New Haven.

Stephen Goodyear, his wife, Mary, their three children, and two servants occupied a large lot fronting on the west side of the green, a location reflecting his status among the wealthiest and most privileged settlers. He had invested one thousand pounds in the venture, and owned 360 acres of land near the town for farming. Like most Utopians, the group planned to live without laws, rules, or jails, governed only by their mutual desire for harmony. Within a year, this ideal had been modified to accommodate reality. A local government was formed with the authority to regulate, tax, and imprison.

The colonists elected Stephen Goodyear magistrate, and later, deputy governor. They prized his impartiality and independent judgment. He heard slander cases, served on civic committees, and was placed in charge of auditing the colony's treasurer. Goodyear presided over the celebrated witchcraft trial of Elizabeth Goodman, who had been behaving oddly and, worse, openly criticizing the Reverend John Davenport's sermons. Residents swore that Mrs. Goodman had cast spells that fouled barrels of beer, made residents sweat and swoon, and even drove chickens to drown themselves in fits of spontaneous madness. Davenport, the powerful leader of the colony, demanded that this woman of "discontented frame of spirit" be severely punished. But

Goodyear's decision was humane, enlightened, and courageous. He ordered no punishment at all, asking only that Mrs. Goodman behave.

To stimulate sluggish trade, the colony leaders settled on a bold plan to build their own trading ship, fill it with local goods, and send it directly to England. The three-masted vessel, built entirely by New Haven craftsmen, was the pride and hope of the colony. They sent the ship out to the North Atlantic almost as soon as it was completed, in the depths of the bitter winter of 1645–46. The cold was so intense that ice thick enough to walk on formed over New Haven Harbor for three miles out to Long Island Sound. Colonists used handsaws to cut a channel to the sea. They loaded the ship with wheat, hides, plate, and other cargo worth a total of five thousand pounds. The ship also carried many of the colony's most prominent citizens. Stephen Goodyear stood on the ice and waved goodbye to his wife, Mary, who was returning to England alone. Margaret Lamberton stood nearby, bidding farewell to her husband, Captain George Lamberton.

Nothing was ever heard of the ship or its passengers again. Not a single identifiable board washed up on any inhabited shore. With the spring thaw came fresh ships from England by way of Boston. None of the crews had seen or heard anything of the ship. The great ship of New Haven, with its passengers and cargo, seemed not to have sunk or wrecked, but to have evaporated. A year and a half after the ship set sail, the passengers and crew were declared dead and their estates settled. What happened next was either the result of mass hysteria or an act of God. On a soft June evening in 1648, townspeople were startled by an apparition hovering over the harbor. First, a keel appeared, and three masts, then sails. Within a few minutes, a fully formed ship hovered over the harbor. The figure of a man stood on the poop deck, left hand on his hip, right hand holding a sword out toward the sea. After hovering for several minutes, the ship suddenly disappeared in a cloud of smoke. Many townspeople swore they had seen the same apparition.

They took it as a sign from God that the ship had gone down and the subjects called to heaven. Two centuries later, Henry Wadsworth Longfellow remembered the event in a poem called "The Phantom Ship," which ended like this:

And the pastor of the village
Gave thanks to God in prayer,
That, to quiet their troubled spirits,
He had sent this Ship of Air.

The loss effectively killed New Haven's hopes of becoming a great trading center. But it indirectly gave one great gift to the world. Stephen Goodyear and Margaret Lamberton would salve their grief and loneliness by marrying. Out of that union came a son, John, the great-great-grandfather of Charles Goodyear.

By the last week of the eighteenth century, New Haven, while no metropolis, had grown steadily to become a bustling port and farming community of five thousand residents. The economy was just beginning to shift from barter to cash. Abraham Bradley, a merchant, offered cash on the barrel for red clover and timothy seed. But at Daniel Read's store, redolent with chocolate, nutmeg, and cinnamon, residents exchanged local butter, cheese, eggs, corn, white beans, flax, and beef bladders for imported goods ranging from Asian spices to European kettles, skillets, caps, and shot. Virtually all manufactured goods above the crudest, makeshift variety had to be imported.

The town approached the final week of the century with little fanfare. Local newspapers offered few retrospectives about the century's being laid to rest, and no mention of Christmas. Timothy Dwight, president of the one-hundred-year-old Yale College, announced perfunctorily that winter vacation would begin January 14; Henry Grilley offered a reward for the return of his stray stallion colt; and the Medical Society of New Haven set its regular meeting for the following week at the Coffee House. At their little house in Oyster Point, a spit of land just south of the town, Amasa and Cynthia Goodyear had more a exciting event to contemplate—the birth on December 29, 1800, of their first child, a son, Charles.

Amasa Goodyear, like his country, was just shy of thirty years old and brimming with ambition and confidence when Charles arrived. An incurable dreamer and tinkerer, Amasa Goodyear had two signature traits, both of which he would pass on to his first son: an ability to see possibilities where others did not, and a debilitating lack of business sense. The latter trait effectively prevented Amasa Goodyear from realizing financial benefits from the former. Following the lead of his ancestor Stephen, Amasa spent several years trying to establish himself in trade with the West Indies. He did this until Charles was five years old. On January 24, 1805, he placed an advertisement in the *Connecticut Journal* inviting shareholders of the Wharf Company to a meeting at his house. Little is known about the Wharf Company (presumably Amasa was its founder), but the news presented to shareholders couldn't have been promising. Within a few months Amasa abandoned the shipping business altogether. His restless imagination turned to manufacturing, and to areas north of New Haven, in Connecticut's rocky, hilly interior.

Until the late eighteenth century, the state's interior attracted mainly trappers, adventurers, and a few other hardy settlers. But with the first tentative flowering of manufacturing in America, this wild region became increasingly attractive for its abundant water power to turn mill wheels. The Naugatuck, a wide, fast river, sliced through a long, gulchlike valley, tumbling over shallow rocks before mellowing into a placid stream near New Haven. Early American manufacturers usually concentrated on small, compact items, giving rise to the term "Yankee notions." The difficulty in getting raw materials to an inland mill was exceeded only by the difficulty and expense of distributing finished products. Some goods made in Naugatuck could be sent in bulk to Hartford or New Haven, but most were carried to the hinterlands through a distribution system as crude as it was quaint. Each spring, peddlers laden with sacks of locally made wares ventured on foot, often for months at a time, stopping in at each lonely house in the woods. So manufacturers concentrated on buttons, combs, cutlery, and buckles. In Waterbury in 1790, a man named Henry Grilley built a mill for making pewter buttons. A decade later, brothers Abel and Levi Porter started making brass buttons. Their success spurred Amasa to try his hand in the button business. With several partners, Amasa purchased a site along Fulling Mill Brook, a tributary of the Naugatuck.

In 1805, just shy of Charles's fifth birthday, the young family packed up and left New Haven on a thirteen-mile journey to their new home. By now, there were two more sons, Henry and Robert, and a newborn daughter, Harriet. They had a good, clean road on which to travel. The Straitsville Turnpike, connecting New Haven and Litchfield, a trading center in northern Connecticut, had opened in 1801. The new road cut the trip from New Haven to Fulling Mill Brook from a day and a half to a relatively smooth half day. But the short distance in miles belied the dramatic change in atmosphere for the young family, from the relative sophistication of New Haven to the rugged, frontier flavor of an area so sparsely settled that it would not officially become the town of Naugatuck for another thirty-nine years. As a coastal trading city, New Haven was replenished at regular intervals by culture, people, and news from Europe and the West Indies, and from other American cities. In New Haven, by the early 1800s, one could buy cognac from France, linens from Ireland, china from Liverpool and gin from Amsterdam. The local newspapers carried dispatches from London, New York, Paris, and Rome. New Haven was home to a fleet of twenty ships known as the China fleet. The ships ventured to the South Atlantic to hunt seals, whose furs they transported to the Far East. The ships returned from the Orient months later bearing exotic fabrics, spices, and tea. The presence of Yale College added a tone of learning and culture that rippled throughout the town. In Naugatuck, cultural attractions were few and sparse. Before a bowling alley opened in 1811, the primary entertainment (outside of a couple of illegal gambling operations) was the daily arrival of the New Haven stagecoach. It was an isolated community of self-reliant people. By the end of the century, thanks in large part to an occupant of the Goodyears' northbound wagon, Naugatuck would be transformed into a major manufacturing center, drawing immigrant labor from around the world. The catalyst would be not the thirty-three-year-old dreamer driving the wagon, but the five-year-old son staring wide-eyed from the back.

The family settled into a small clapboard saltbox cottage on the east side of the Naugatuck River, near the confluence of the Naugatuck and Fulling Mill Brook, an area known as Union City. The cottage was dominated by a central fireplace and chimney that rose from the center of the house and supplied heat to each room. While Cynthia set up house, Amasa set up the mill.

Like most early manufacturers, indeed, like most Americans, the Goodyears raised crops for food, to supplement business income. Charles, the oldest child, bore a large share of the farm chores. Despite its crude cultural offerings, Naugatuck had its charms, most of them outdoors. Most boys hunted partridge and pheasant in the woods, fished for trout, skated on mill ponds in the winter, and plunged into swimming holes in the brief, hot summers. But Charles was a sickly, serious boy who had little use for games or physical activities outside of farming. Frail from early childhood, he suffered all his life from digestive ailments that were classified under the catch-all term "dyspepsia." He preferred to spend his time reading. From the time he was ten or eleven, neighbors remarked on his adultlike seriousness. Nor was he mechanically inclined—a surprising fact considering that he would make his mark as an inventor. He was actually something of a klutz. In his memoirs, Goodyear confessed to "an aversion to bestowing thought upon machinery where there is anything complicated about it."

Charles also spent a good deal of time at his father's mill. Despite his lack of interest in mechanics, he was fascinated by the creative process and the excitement of new ideas being turned into products. For a few years at least, Amasa's business did fairly well. By 1807, he was making the first pearl buttons manufactured in the colonies. In 1812, American troops marched off to battle the British wearing coats fastened by buttons turned at the Goodyear mill. Before long, Amasa expanded his products to include spoons, scythes, and other small tools. He had watched local farmers straining their backs lifting hay and manure with clunky, heavy, imported iron pitchforks. He got the idea for a lighter-weight fork with steel tines, which he called "the spring steel hay and manure fork." His proud son years later would declare the forks to have been "universally considered in the United States one of the greatest improvements ever made in farming implements." But tradition-bound farmers were slow to warm to the device. Amasa had to give several away as samples before they started to catch on.

The Goodyear children attended the local school. But when Charles was sixteen, Amasa, during a period of apparent prosperity, hired a young man named William DeForest to tutor the children. It was an important meeting for all concerned. DeForest would develop into a successful woolen manufacturer and marry Charles's sister, Harriet. He would also become one of Charles's most steadfast supporters, an angel on his shoulder during his darkest years. Later DeForest would become an important partner in the Naugatuck rubber works that evolved after Goodyear's discovery. As a tutor, DeForest noticed and later remarked on Charles's seriousness and maturity, and his eagerness to learn. From an early age, Charles was intensely religious and a self-assigned student of the Bible. If most churchgoing parents have to cajole their children to stop fidgeting and pay attention, Amasa and Cynthia faced the opposite problem with young Charles. Amasa especially grew concerned about Charles's attachment to the Congregational Church, and the boy's budding interest in becoming a minister. Amasa had other plans for his firstborn son. By the time Charles was in his teens, he was more than just a helper in his father's mill, he was a junior partner. Amasa could ill afford to lose a source of serious, industrious—not to mention free—labor, not even to the church.

Perhaps with the idea of driving a wedge between Charles and his church plans, and also giving his son some practical business experience, Amasa packed the boy off to Philadelphia at the age of seventeen to learn the hardware trade. Charles would spend the next four years as an apprentice to the firm of Rogers & Brothers, then one of the leading hardware importers in the country.

By all appearances, Goodyear worked hard and enjoyed himself enough that for a time he considered going into the importing business himself. Toward the end of his stay, though, illness forced him back to Naugatuck. The only clue to the nature of his sickness comes from Goodyear's own memoirs, written throughout in an awkward third-person voice that is elliptical and cryptic in revealing personal details: "By close application and hard labor in this business, his health became much impaired, so that at the expiration of his apprenticeship, he was greatly disappointed by being obliged to abandon the idea of establishing himself in the business he had designed to pursue." It's hard to imagine a healthy lad of eighteen or nineteen falling ill due to overwork at a hardware store. But then, Goodyear had never been a healthy lad. After four years he found himself back in Naugatuck, working with his father at a firm now called A. Goodyear & Son. In addition to its existing line of hay and manure forks, table forks, spoons, and tools, the company now manufactured clocks. After recuperating from his ordeal in Philadelphia, Goodyear spent the next five years working with his father.

Some time after returning to Naugatuck Goodyear began exchanging furtive glances with a fellow parishioner at the Congregational Church. Clarissa Beecher was a pretty, dark-haired girl with large, brown eyes. Clarissa's father, Daniel Beecher, operated an inn on the west side of the Naugatuck River. It was an impressive white building, especially ornate for the time and location, with a broad front porch and a Palladian window in the center, and stables to the side. The inn was strategically located adjacent to the toll house for the Straitsville Turnpike—a natural resting place for weary travelers. The business prospered and the Beechers were one of the town's leading families. Daniel Beecher donated the land for the village green that to this day anchors Naugatuck's downtown.

Clarissa Beecher was obviously a good catch for Charles Goodyear. At the time, Goodyear must have seemed like a good catch for Clarissa. Although Amasa Goodyear had never managed to make a great deal of money, he had kept his business going longer than the majority of entrepreneurs—he had been a Naugatuck manufacturer for nearly twenty years. The spring steel hay and manure forks were selling well. There was optimistic talk about branching out in hardware sales elsewhere in the country. Daniel Beecher wasn't the only local businessman to donate land to the town. Amasa contributed a plot as the site of an early schoolhouse. A wing of Naugatuck High School today carries the title Goodyear House, for Amasa rather than Charles, to commemorate his gift. Charles, though slender to the point of being gaunt, was a rather handsome young man, with thick, dark hair; a long, straight nose; full lips; and deep-set eyes under a dark brow. More important, at least to Daniel Beecher, Charles gave every appearance of being a man who would make his mark on the world. He was serious, industrious, clever, fairly well-educated, and God-fearing. His hard work on behalf of his father's business, his apprenticeship in Philadelphia, and the obvious weight Amasa placed in his counsel, could not have helped but impress a potential father-in-law.

Had Daniel Beecher known the hardships that being married to Charles Goodyear would impose on his daughter, he might well have sent her off to live with relatives until the love fevers cooled. Offered a glimpse of the future, Clarissa herself might have fled willingly. Charles Goodyear would, indeed, make his mark on the world, but at a huge cost to himself and everyone closest to him. His obsession permitted him to endure almost any amount of suffering and to tolerate suffering among those he loved. The glory of his mission was so obvious to him that suffering, if anything, became a sacrament. He sensed that his quest would yield a discovery to change the world. The fact that he was right hardly lessened the tremendous burdens he placed on his family, especially Clarissa. In the end, she gave him a degree of loyalty that he had no right to expect.

The great statesman Daniel Webster, arguing for Goodyear in a court case decades later, would say of Clarissa: "In all his distress, and in all his trials, his wife was willing to participate in his sufferings, and endure everything, and hope everything; she was willing to be poor; she was willing to go to prison; she was willing to share with him everything, and that was his solace. There is nothing upon earth to compare with the faithful attachment of a wife." Of course, the now-obvious question is, what was in it for Clarissa, what was to be *her* solace? But few asked such questions in the early nineteenth century. Even in an age when the suffering, silent, loyal wife was the model for a male-dominated world, Clarissa's suffering loomed large.

Contemporary descriptions of Clarissa tend to hagiography. The Reverend Bradford K. Peirce's description in a biography of Charles published six years after his death (and thirteen after Clarissa's) is fairly typical: "She was . . . of a peculiarly amiable temper, endowed with great fortitude, and sustained by a sincere religious faith. She was devoted to her husband, and endured without a murmur the extraordinary sacrifices occasioned by his long-continued and unsuccessful efforts, before he made his great discovery. Although yielding in temper, she had a strong hold upon the confidence of her husband, and was his constant confidante and counselor. In the heat of his fervent nature, when suddenly resolving upon some new measure, or giving way to the most sanguine expectations, she would place her hand quietly upon his shoulder and say, 'Do not make up your mind too hastily; had you not better wait until tomorrow?' The touch of that wise and gentle hand was always effectual. How faithfully, through all the years of his trial, this most excellent Christian woman supported her husband, aiding with her own hands in his unremitting experiments, and sympathizing with all his fluctuating fortunes, can only be known to those who were the constant observers of her beautiful life."

It is difficult to fault the Reverend Peirce for writing of Clarissa in this way. Biographies tend to be worshipful. And Clarissa's devotion fit a neat model; one needed not worry about Clarissa after writing her off as a saint. And yet one longs for a crack at those "constant observers of her beautiful life" that Peirce mentioned, the living relatives and acquaintances whom he interviewed for his book. No human being could have borne up under the pressures Clarissa did without blowing a gasket from time to time, without questioning and requestioning her devotion to this man who wouldn't stop playing with rubber even as their lives crumbled around them. The pressures on this most excellent Christian woman amounted to far more than just material privations, and those were dire enough. The nature of Goodyear's experiments made them particularly intrusive even in those areas where a woman of the early nineteenth century usually held unquestioned dominance.

During the years when Goodyear could not afford his own laboratory, his home substituted. The best, and often only, source of heat for his experiments was Clarissa's kitchen stove. It is not clear which times were more stressful for Clarissa, the months at a stretch when he left his family behind with little or no money in order to pursue experiments in some distant town, or the times when he was home, conducting experiments that always seemed to involve a substance or compound stickier and smellier than the last.

Even a small chunk of raw rubber sitting on a desk exudes a sharp vegetable odor, pungent and sickly sweet. But slicing, chopping, or heating rubber has the effect of releasing an odoriferous genie from a bottle. The smell infests hair, clothing, furniture. And rubber was just the base ingredient. Goodyear mixed it with magnesium, turpentine, lime, lampblack, sulfur, nitric acid—a panoply of offensive (and sometimes lethal) substances. Clarissa's kitchen (indeed, the entire house, since quarters were most often cramped) usually reeked of Charles's latest experiment. Splotches of rubberized goo clung to her pots and pans, doubling as scientific vessels in the wee hours. Through it all, she struggled to hold her home together, to keep her children clothed and fed. There is no denying that Clarissa Goodyear was a remarkable woman—the mere fact of her staying with Charles under such conditions for so many years qualifies her for that description. Unfortunately, the too-few scraps of historical evidence and hagiographic testimonials leave us only to guess at her unavoidable frustrations and the source of her magnificent forbearance. Clarissa remains as much a mystery to us today as her own future life of hardship must have been to her on August 27, 1824, three days after her twentieth birthday, when Daniel Beecher gave his daughter's hand to Charles Goodyear.

2

A HIGH PURPOSE

The first years of Charles and Clarissa's marriage were good ones, full of bright prospects and promise. Charles enjoyed a period of good health. The steel hay forks had caught on with farmers. The success emboldened Amasa and Charles to look beyond the provincial confines of Naugatuck and a life of peddling Yankee notions to local hayseeds. In 1826, they decided to open a hardware store in Philadelphia, capitalizing on Charles's experience and contacts there. Amasa would continue to run the manufacturing operation in Naugatuck, while Charles would set up and run the store. Charles and Clarissa left that year for Philadelphia. Goodyear rented a shop on Church Alley, connecting Second and Third Streets, just around the block from his old employers, Rogers & Brothers.

The Goodyear shop was well-stocked with the hay forks from Naugatuck, plus axes, knife sharpeners, Verre & Company's mill-cut and small saws; Carrington's coffee mills; Scovill's gilt buttons; Hive Works's metal and ivory buttons, calf and morocco pocketbooks and wallets; John Yard's bone suspenders and shirt buttons; Ezra Williams's ivory combs; Howard Pratt's wood combs; L. P. Lee's fine plated ware; D. B. Thompson's shaving brushes; Mix's Britannic spoons; N. C. Sanford's patent augurs; and Globe Company's scythe stones. Despite its out-of-the-way sounding name, Church Alley was a busy, well-traveled row of shops and offices. A few doors down from A. Goodyear & Son, Dr. Cornwell offered general medicine and surgery, as well as treatment for impotence and "a certain complaint incident to both sexes"—presumably syphilis. Dr. Cornwell assured his patients that he had cured thousands of people of this ailment while serving as a military doctor, and promised a cure "without an hour's detention from business or change of diet, or a possible discovery by the most intimate friend."

Unlike Rogers & Brothers, Goodyear's old employer, A. Goodyear & Son limited its stock to domestically produced items. It was a bold statement of faith in the budding crop of American manufacturers, at a time when virtually all store-bought goods of any sophistication were still imported, mainly from England. Established Philadelphia merchants hardly considered Goodyear a threat. Most, if they acknowledged the company at all, predicted a quick demise for merchants foolish enough to restrict themselves to American-made goods—whose reputation was either unenviable or nonexistent.

The Goodyears manufactured much of the merchandise themselves. Amasa shipped scythes and forks, almost certainly by sea, in rough form, to Philadelphia, where Charles finished them for sale. The shop initially prospered. The young family for a time enjoyed a steady income and material comfort. Clarissa had given birth to a daughter, Ellen, in July 1825, a year before the couple moved to Philadelphia. During this happy period they added a second daughter, Cynthia, in 1827, and a third, Sarah Beecher, in 1830. By 1829, Goodyear had established himself as a solid member of the Philadelphia merchant community. His advertisements appeared prominently in the *Aurora & Pennsylvania Gazette*, a daily newspaper, alongside those for daily steamship lines to New York, a piano warehouse, and the Reverend T. Bartholomew's Elixir of Life (shown to cure ague in a single day, bloody flux and dysentery within hours).

"A handsome fortune was accumulated by the firm, and the writer occupied a position in business every way desirable," Goodyear recalled with a note of pride in his memoirs. Goodyear was a devoted husband and father during these years; even afterward, when life became more cruel, he never sought refuge in the conventional sins. He was an occasional drinker and smoker until, during a business trip to New York in 1826, the sight of drunks spilling from the taverns into the streets shocked him into lifelong abstinence. "I have quit smoking, chewing and drinking, all in one day," he happily reported in a letter to Clarissa. There is no indication that he ever sought refuge from his many troubles in the bottle.

In 1830, though, the prosperity that had seemed so solid suddenly evaporated. Early success had prompted Charles to extend operations in many states from his Philadelphia base. He shipped goods around the Mid-Atlantic and into the South to buyers who often had flimsy credit. A punishing bill called the Tariff Act of 1828 set artificially high prices on goods manufactured in New England. When the economy bottomed out, many of Goodyear's customers simply quit paying. This in turn left Goodyear with mounting debts. When creditors came calling, he pleaded for extensions, which allowed him to continue operating for a time. But new creditors bought the debt from the original ones and soon Goodyear found himself pleading with creditors less forgiving than the old. Goodyear considered declaring bankruptcy, but that would have required him and Amasa to hand over the entire company, including all inventory, the Naugatuck mill, and rights to any patented inventions. Goodyear moved from his Church Alley shop into smaller and less expensive quarters on the same street.

He later recalled his decision to hang on in business a mistake. "The writer did not count upon the disadvantages he had to contend with, on account of impaired credit, and did not know, what experience has since taught him, that under circumstances of embarrassment, the only wise course is not to continue the same business, at least not to continue it under suspended liabilities."

Stress from his business helped to bring on an attack of dyspepsia so severe that he was unable to leave his bed for days. He was only thirty-four years old. Yet as he writhed in agony, occasionally becoming delirious, Clarissa feared for his life, and for the possibility of being left alone with three small children. Gradually, his health improved. No sooner had he recovered to the point that he could move about than a sheriff arrived at his door early one morning. Daughter Ellen, the eldest at five, woke to discover her father gone and her mother preparing to leave. Clarissa instructed Ellen to look after the other children while she visited their father at the debtors' prison.

In Philadelphia, debtors shared five cells in a dank old prison on Arch Street. The cells were twenty feet square and housed eight to twelve prisoners each. While debtors were kept separate from common criminals, their conditions were hardly better, and in some ways worse. A father of five, suddenly out of a job and unable to pay a grocery bill, might find himself locked up with assorted frauds and cheats, along with alcoholics, shy their tavern bills, writhing on the cell floor in full-blown delirium tremens. Inmates whose families and friends could afford basic provisions ate relatively well and slept on beds. But the poorest debtors had nothing but bread and water. They slept on the floor and scavenged for any food beyond bread. Crowded cells, dirty inmates, dreadful sanitation, poor nutrition, and nonexistent medical care provided an excellent environment for disease, so a term in debtors' prison all too frequently amounted to a sentence of death.

Debtors' prisons represented probably the worst institutional hypocrisy outside of slavery for a young nation professing itself to be dedicated to liberty, justice, and individual rights. Thousands of men and women (though some states forbade imprisoning women for debt) lost their freedom for the most trifling sums. Of 2,000 people cast into debtors' cells in New York City in 1816, more than a third owed less than $25 apiece, more than half less than $50. In Philadelphia in 1830, when Goodyear was first jailed there, 452 inmates languished for debts *totaling* less than $1,500—or, $3.18 per person. The bottom 40 prisoners owed an average of 58 cents, with one owing just 2 cents. Jail terms averaged 20 days, but some terms could drag on for months.

More sinister than the trivial price placed on freedom here in the cradle of liberty was the arbitrary and capricious manner in which terms were meted out. Predictably, the poorest and most defenseless citizens endured the greatest suffering. In tangible ways debtors had fewer rights than violent criminals. Unlike ordinary criminals, who were answerable to the state, debtors essentially became the personal property of their creditors, in lieu of repayment. Creditors decided when and why a debtor was arrested, and, short of repayment, they decided when the debtor would be released. In Philadelphia, creditors paid a "bread fee" of twenty cents per day, for a loaf and two blankets to keep their debtors from starving and freezing. Since the bread fee often exceeded the amount of the debt, and since prisoners had no way of earning money to repay while incarcerated, an extended term often served merely to satisfy the creditor's desire for revenge. Debt prison terms were often called "spite actions." Release was as arbitrary as arrest. In many cases, a debtor simply remained in jail until the creditor grew tired of paying the bread fee.

The physical suffering was nothing compared with the despair and frustration of dealing with a court system that might have served as a model for one of Kafka's twentieth-century European nightmare bureaucracies. In Philadelphia, one inmate's Byzantine odyssey began when he had the audacity to sue an employer for five dollars in unpaid wages. The court sided with the employer, not surprising since the employee was described as "a poor and ignorant black man." Unpaid wages quickly became the least of the man's worries. Having triumphed in court, the vengeful employer now demanded fifty cents to cover his court costs. Not to be outdone, the court demanded two dollars for its troubles. Because he had not been paid for his work, the luckless worker had no money to pay off these new debts, and was promptly imprisoned at Arch Street. Black and white debtors stayed in segregated cells, but white debtors could hardly expect better treatment from the courts. In New York in 1811, one aging veteran of the Revolutionary War found himself locked up for months over a debt of three dollars.

Reformers in the United States and Europe had been trying for decades to eradicate debtors' prisons. They would not succeed until the late nineteenth century. Unfortunately for Charles Goodyear, debtors' prisons were still going strong when A. Goodyear & Son ran into trouble. He was released after a short time, following this first arrest, but this would not be his last encounter with debtors' prison. "These trials were not wholly without their advantage; lessons of life were learned from them," he insisted in his memoirs. "If anyone is desirous to learn more of human nature than he can learn in any other way, or wishes for a moment to look upon the darkest side of life's fleeting shade, let him, for such a cause as debt and misfortune, be placed within the bars of a prison door, without a dollar in his pocket, and in conscious innocence look out upon the world, and reflect upon the wide contrast in his condition with that of those who are enjoying liberty without; while within he finds his fellow sufferers all upon the same level, whether incarcerated for the sum of one hundred pence, or of one hundred thousand pounds. Then, notwithstanding the mortification attending such a trial, if he has (as every human being should have), a good purpose in life for which to live and 'hope on,' he may add firmness to hope, and derive lasting advantage by having proved to himself, that, with a clear conscience and a high purpose, a man may be happy within prison walls, as well as in any other (even the most fortunate) circumstances in life."

Goodyear would need that positive outlook outside prison walls as well as in. Stephen Bateman, Goodyear's cousin on his mother's side, moved to Philadelphia from Naugatuck in 1831 to help Charles try to salvage the store. But even with the extra manpower, Goodyear finally recognized the futility of continuing. He declared bankruptcy, surrendering the rights to his store and, with it, to the inventions of A. Goodyear & Son, including the rights to the steel forks, to his creditors. Even this capitulation could not keep all of the creditors away. Those who felt slighted in the initial bankruptcy hounded him for years.

In the course of a few months Goodyear had gone from being a prosperous young Philadelphia merchant to a debtor and jailbird with few prospects. Family tragedy put a coda on the brief happy years the Goodyears had enjoyed. A daughter, Clarissa, born in 1831, died before her first birthday. New Year's Day 1833 saw the birth of a son, Charles Jr. But the joy of that birth was soon offset by the death in the same year of Sarah Goodyear, at three.

Broke, sick, and grief-stricken, Charles began casting about for some new line of work to support his family. Inventing was not a practical answer, but it was perhaps the natural choice for Amasa Goodyear's son. Charles Goodyear decided to invent his way out of poverty. He worked on improvements for buttons, faucets, and spoons. But he was never able to move these beyond the idea stage.

Until the early 1830s, Goodyear appears to have been largely unaware of rubber or indifferent to it. In his memoirs, he speaks of having come across a piece of rubber as a schoolboy, and being amazed by "the wonderful and mysterious properties of this substance." It's possible that a piece of rubber made its way into Amasa Goodyear's shop when Charles was a boy, but his memory may have been playing tricks on him. Even in England, rubber did not begin making a significant appearance until around 1810. The United States did not start seeing significant quantities until the 1820s.

His introduction to the substance came in 1834, when Goodyear visited New York City and happened past the display window of the Manhattan office of the Roxbury India Rubber Company. Started in Boston, the Roxbury Company was the largest and strongest of the early wave of American rubber companies. Goodyear examined a life preserver in the store. It was not the rubber that attracted Goodyear's attention, but the valve used to inflate the preserver. Certain that he could devise a better valve, Goodyear bought the preserver and took it home with him to Philadelphia. He worked for weeks developing a more efficient valve. Had he begun his experiments a year or two earlier, he might have sold the device to the Roxbury officers for a decent profit. As it happened, his timing was off. When Goodyear returned to the New York store where he had bought the item, the agent examined the valve for a few moments, praising the ingenuity. Then he shook his head sadly.

"I want to show you something," the agent said. He led Goodyear to the warehouse behind the showroom. He pointed at rows of shelves containing heaps of misshapen blobs, their folds stuck fast together. The room was pungent and foul-smelling. There would be no use for Goodyear's improved valve. The rubber industry, the agent explained, was in deep trouble.

The agent paused, then uttered a few words that would change Goodyear's life, and history. "Forget your valve," he said. "The one invention that could save the rubber industry and earn a fortune for the inventor would be a method for preventing rubber from . . . this." An inventor who could pull that off would make his mark on the world. The words rang in Goodyear's head after he left the shop for home.

The American obsession with rubber during the early 1830s bears a striking resemblance to the Internet frenzy of the 1990s. In each instance a new industry emerged, one so promising and so unlike any before it that investors and entrepreneurs alike willingly blinded themselves to pitfalls, even ones that in retrospect seem obvious. Just as speculators drove the value of untested, unprofitable dot com start-up companies to surreal levels during the 1990s, thousands of nineteenth-century New Englanders poured their savings into the miracle substance called rubber without knowing much of anything about it. In both cases, the education was hard-earned and painful.

Rubber comes from latex, a milky white fluid produced by about three hundred varieties of trees, vines, and shrubs around the world. Though it is most often associated with the tropics, you can find latex in such everyday plants as goldenrod and dandelions. The Soviet military, faced with rubber shortages in World War II, actually experimented with cultivating vast fields of a dandelion relative to supply their needs. Decades earlier, Thomas Edison planted thousands of acres of goldenrod in the southwestern United States, hoping for a cheap and plentiful source of rubber for his booming electrical industry. But the poor quality of the rubber and the meager supply per plant killed both efforts.

The single plant favored above all else, the one that came to be known simply as the "rubber tree," is the species *Hevea brasiliensis.* These trees today occupy vast plantations in Asia, but in the early days of rubber, the only source was the jungles of the Para region of northern Brazil. Gathering the latex in the dense jungles was an arduous task. Often referred to mistakenly as sap, latex is actually produced not in the core of the plant but between the outer bark layer and the inner soft-bark layer, or cortex. Latex flows best in the early mornings, so workers would venture out around dawn, making lateral, angled slashes in the outer bark, and affixing a clay cup near the bottom of the incision. By varying the location of the incisions, workers could avoid hurting the tree. A *Hevea brasiliensis* tree, growing to sixty feet in height, might give latex for twenty-five years.

One speaks guardedly of the European "discovery" of rubber, because the native population of Central and South America had been using it for ages before the first European explorers set foot in the New

World. At the London Science Museum you will see an object enclosed in glass that is at once exotic and heartbreakingly domestic—an ancient rubber ball, found in the grave of a Peruvian child. The first recorded instance of European exposure to rubber dates to the second voyage of Christopher Columbus, during the 1490s. Michele de Cuneo, a member of the party, reported seeing trees that "give milk when cut," and that the native people "made a kind of wax." A couple of decades later, Hernando Cortés was said to have observed a game involving a bouncing ball at the court of Montezuma II. Throughout the region, people played an elaborate game called "batey," a sort of forerunner to basketball or hockey, in which teams tried to get a rubber ball through a circular stone with a hole cut in the center.

Although explorers took samples home as curiosities, Europeans did not investigate rubber in earnest until the eighteenth century. In 1736, Charles-Marie de la Condamine, a French naturalist, traveled to Peru and began an extended odyssey into the countryside. Condamine noticed the natives collecting "a white resin like milk" from incisions at the foot of various trees. He reported that the natives made boots and bottles, which they hardened by smoking in layers over a fire. Condamine himself coated pieces of canvas with the latex, which, when held over the smoke, produced waterproof fabric. Condamine called the substance caoutchouc, a rough approximation of the native term meaning "weeping wood."

In England, the first practical use for caoutchouc—pronounced *cow-CHOOK*—was decidedly humble. Stationers had discovered the material handy for rubbing pencil from paper. Joseph Priestley, an eminent eighteenth-century scientist, cited the development in the introduction to a scientific paper: "I have seen a substance excellently adapted to the purpose of wiping off from paper the marks of black lead pencil. It must therefore be of singular use to those who practice drawing." In terms of manufacturing, it was an uninspiring debut for a substance of such limitless potential. But the development yielded one important thing: the term "rubber."

Raw latex is often compared with milk, not just for its consistency but for its tendency to spoil quickly when exposed to air. Latex was too delicate for the long voyage overseas. For export, rubber workers smoked latex over coconut fires to remove the water. What arrived in Western ports were hard, knobby lumps that were difficult to work with. In the 1760s, French scientist François Fresnau discovered that rubber would dissolve in turpentine. The resulting liquid could be spread on cloth to create waterproof fabric. Left to dry, the turpentine would eventually evaporate from the surface. The evaporation was never perfect—the rubber remained tacky, especially in warm weather—but it was a significant breakthrough nonetheless. During the late eighteenth century, French balloonists had some luck coating silk balloons with rubber to make them airtight.

The first rubber goods in the United States were crude shoes made at the site of the rubber trees in South America. Workers smoked layers of latex over wooden lasts. In 1826, New England imported 8 tons of rubber shoes from Brazil. In 1830, citizens snapped up 161 tons. Inevitably, American entrepreneurs caught on to the idea of manufacturing their own products from rubber.

The Roxbury India Rubber Company was among the first, in 1833. A founder, Edwin M. Chaffee, a former leather worker, had become enamored of the substance after experimenting with ways to make leather more waterproof. From an initial investment of $30,000, capitalization mushroomed to $240,000 a year after the company was founded. By 1836, investments were tipping the coffers at $400,000. The Roxbury Company was the first and the biggest of the companies caught up in what came to be known as rubber fever, but it was by no means the only one. Other hopefuls sprouted in Boston, Chelsea, Framingham, Salem, and Lynn, Massachusetts, as well as in Connecticut, New York, and New Jersey.

When the first really hot summer came along, everything changed. One of the more spectacular product failures was visited on John Charles Frémont, the famed Western explorer. Frémont had ordered an air-filled rubber boat for a trip that would take his party to the Great Salt Lake. When the boat was delivered to his home in Washington, the aroma was such that the boat had to be moved from the house to the stable, and the house sprayed with disinfectants. The boat served Frémont's party well enough until they reached the Great Salt Lake. All at once, the vessel simply began to come apart. Only furious rowing and constant bellowing into the air compartments got Frémont and his men to shore. The boat had been made by a New Jersey rubber manufacturer named Horace Day. More about him later.

Thousand of customers were experiencing similar if less spectacular problems. Life preservers, coats, shoes, and hats left too long in the sun turned sticky, then gooey and shapeless. All early rubber products had a pungent smell. When they began to melt, the smell was atrocious. Bewildered manufacturers and store owners threw up their hands in dismay. The Roxbury India Rubber Company took back twenty thousand dollars' worth of merchandise in 1834 alone.

Goodyear began his first crude experiments upon returning to Philadelphia from New York. Fortunately for him, the industry's decline made the stuff plentiful and cheap. You could buy raw rubber from Brazil for a few pennies a pound. With no guidebooks or experts to imitate, and no chemical knowledge or training, Goodyear fumbled along at first in his experiments according to the oldest and most enduring scientific method—trial and error. The one well-known rule, thanks to François Fresnau, was that turpentine could reduce hard rubber to liquid. It was the starting point for any neophyte such as Goodyear. But before he could mix the rubber, he had to painstakingly break it down, first by tearing away small chunks, then rolling or kneading it by hand until soft. It was monotonous, finger-numbing work, especially if, like Goodyear, your only tools were your fingers, a table, and your wife's rolling pin.

Shortly after his return from New York, Goodyear was once again arrested and jailed for a debt stemming from the hardware business. He brought with him to his cell some samples of rubber and a rolling pin he borrowed from Clarissa. Some of his earliest experiments took place in these most humble and humiliating circumstances, with Goodyear, no doubt much to the amusement of his fellow inmates, patiently rolling bits of raw rubber into a malleable consistency. With the hardware business gone, broken up and sold off in bankruptcy, and with Goodyear ignoring all other pursuits as he dove into rubber, the family began to feel the poverty that would engulf them for years.

The family had moved from Church Alley into a small home on Race Street in Philadelphia. When the need called, Goodyear took family possessions off to pawnshops, a practice he would continue for the next decade or so. Among the items to go were pieces of furniture and silverware. He spared a set of china teacups, not out of sentiment but because they could double in the evenings as mixing bowls for rubber and turpentine. With little to guide him, Goodyear selected substances to mix with rubber almost at random, hoping something good would happen. It was a sloppy and haphazard way to conduct experiments, to be sure. However, with so little known about the chemical properties of rubber, Goodyear's methodology was not much different from that of other inventors or manufacturers.

In his early experiments, one substance that showed promise was magnesia, which he mixed with turpentine and rubber. Magnesia made the surface of the rubber smooth and also turned the mixture white. Most rubber products were a dreary, dingy greenish brown. Goodyear, a romantic at heart and easily moved by aesthetics, was as pleased by the white appearance as by the apparent loss of tackiness. Delighted, he began to make objects out of the mixture. One of the first was a purse with a steel clasp, for Ellen to take to school. For a brief time, he believed that he had unlocked the key to rubber's mystery. He was mistaken.

3

FALSE GLORIES

A man intent on devoting his life to invention could scarcely have picked a more fertile and promising time in which to live, nor one more ripe with the potential for personal disaster. The story of pre–twentieth-century America is usually dominated by the twin pillars of the Revolution and the Civil War. And yet the decades in between—neatly encompassing the life of Charles Goodyear—were in themselves a period of extraordinary creativity. It fell to these Americans to forge a country out of the grand ideas handed down by the Founding Fathers. They had incredible tools at their disposal: a vast, underpopulated land brimming with natural resources, and (for white Americans, at least) unprecedented liberty and opportunity to exploit them. Thomas Low Nichols, a New Hampshire physician who grew up during the early 1800s, wrote of his fellow countrymen: "The Yankee's first thought of Niagara is the number of water wheels it would turn, and the idea of its sublimity is lost in sorrow at the terrible waste of motive power."

If this tendency to see the natural world as a manufacturing opportunity had its dark side in the clear-cutting of ancient forests and the befouling of waterways, it is also true that the decades between the Revolution and the Civil War unleashed a productive energy that transformed the United States from making Yankee notions at the start of the nineteenth century to matching and overtaking England as the industrial engine of the world by the end of it. Alexis de Tocqueville, not yet thirty years old when he made his famous tromp across America in the early 1830s, was alternately captivated and appalled by a civilization "without parallel in the history of the world."

"The whole community is simultaneously involved in productive industry and commerce," de Tocqueville wrote. And while the Frenchman lamented that Americans' obsession with business left little room for intellectual thought or high art, he was plainly astonished at the profusion of practical inventions that issued from the American mind. Goodyear's home state of Connecticut, in particular, teemed with inventors. Eli Whitney, from a small town near New Haven, invented the cotton gin in 1793. The machine, which separated cotton fiber from seeds, replaced painstaking manual labor and practically overnight converted cotton from a luxury fabric to a material for the multitudes. Later Whitney, along with Simeon North in Berlin, Connecticut, pioneered the use of interchangeable parts in gun-making. Interchangeable parts changed the face of manufacturing forever, by replacing the piece-by-piece craftsmanship of artisans with mass production.

By mid-century, American life had already begun to shift away from the rural and agricultural toward the urban and industrial. In 1850, the value of American manufactured goods topped $1 billion, surpassing the value of agricultural products for the first time. And while most Americans still lived on farms or in small towns, 20 percent now lived in cities of 2,500 or more, up from 6 percent in 1830. And each of these cities teemed with inventors and entrepreneurs. But the freedom to dream, invent, and build carried with it the freedom to fail spectacularly. A classless society extended few safety nets. John Fitch, a pioneer in river steamboats, foretold the fate of many inventors to come when his dream of a steamboat empire crumbled into financial failure and suicide in 1798. In the end, what separated Charles

Goodyear from many other inventors was his almost superhuman capacity to endure.

After eight progressively harder years in Philadelphia, the Goodyears had little reason to remain. The city that had seemed to hold such promise for them was now etched with the memories of two dead children, a failed business, and intimate acquaintance with one of the city's most unsavory accommodations, the debtors' prison. Poor and homesick, they returned to Connecticut in 1836. They did not travel as far as Naugatuck, where Amasa's mill had closed down along with the Philadelphia hardware store. Instead, they stopped in the city of Goodyear's birth and ancestry, New Haven.

There would be no choice lot on the west side of the green for this descendant of Stephen Goodyear. Charles found instead a small cottage on Congress Avenue, one of the main approaches to the city from the southwest. Congress Avenue today is a mundane stretch of frame clapboard houses interspersed with squat brick buildings, some empty and festooned with graffiti. History suggests that the street has improved since Goodyear's day. An early planner optimistically named the area around Congress Avenue Mt. Pleasant, but the name that stuck was Sodom Hill, which more accurately reflected the variety of illicit pleasures that could be found there. Geographically, the "hill" was little more than a barely perceptible hump protruding from marshes that cut the neighborhood off from the rest of the town and made it especially susceptible to plagues. The free blacks and poor whites who occupied the shanties and simple frame cottages on Sodom Hill found work at the nearby tanneries and dockyards. The Goodyears occupied a cottage known as the Sodom House, a name fraught with irony for a man who saw himself as fulfilling a mission from God. Probably it fed his sense of noble suffering.

Passion, beyond a certain level, cannot be faked. When you encounter true passion, you cannot help but be impressed, moved, excited. Whatever else one may say about Charles Goodyear's financial or organizational abilities, he was always authentic on the subject of rubber. From the time of his first experiments he was unreservedly, wholeheartedly passionate about the stuff. He believed. This made him an irresistible salesman when he needed money. When times grew darkest he was always able to convince some neophyte that a breakthrough was just around the corner. Goodyear may have been mistaken, but he was not lying; in his mind Goodyear stood forever on the precipice of success.

The first of his investors was Ralph B. Steele of New Haven, about whom little is known except that he fronted Goodyear enough money to cover his family's modest expenses and to continue his experiments for a time. Goodyear sold Steele on the potential of his magnesia formula, which he continued to work on in New Haven. There was no question of affording a separate shop, so the cottage that served as home to Charles, Clarissa, Ellen, Cynthia, Charles Jr., and William, born in May 1836, doubled as a laboratory. From their earliest memories the children lived in a rubber-coated world. Ellen would later recall the cottage littered with scraps from various experiments. Strips of rubber-coated fabric left to dry clung to windows like rain-soaked leaves. Every saucer, cup, and plate contained another batch. The fetid, sickly sweet stench of raw rubber and the nostril-clearing sting of turpentine permeated the house and crept into clothes, hair, and skin. As they grew, the children were impounded into labor around the house, helping fashion crude rubber shoes, aprons, caps. After a while, Goodyear built a rough shed next to a fence separating his own yard from a neighbor's, and moved his experiments out there. Neighbors complained about the smell. Even Sodom Hill had its limits. Whoring, boozing, gambling, and hell-raising were one thing, but rubber was simply too much. To his neighbors, Goodyear was plainly a fool or a madman, more likely both. How else to explain a man working so hard surrounded by an ever present putrid funk, earning nothing but deeper poverty as his reward?

Some of his magnesia-rubber pieces emerged with their surfaces tantalizingly smooth. Others had the tacky surface that was the bane of most early rubber goods. Goodyear had long suspected that turpentine, the solvent needed to reduce rubber to a mixable state, might be marring his products. Although most of the turpentine evaporated after mixing, Goodyear was frustrated by the residue. Moreover, turpentine was expensive, messy, and time-consuming. A gallon cost around seventy cents, and that much could be used dissolving a single pound of rubber. The rubber first had to be painstakingly cut into small pieces, then stirred every hour or two until it dissolved. Most often he had little choice but to use lump rubber, since that's how it arrived in New Haven. So when Goodyear learned of some rare barrels of latex—rubber still in its liquid state—for sale by an importer, he leapt at the opportunity. The latex wasn't pure. It had been combined with alcohol to prevent spoilage. But at least he'd have the chance to test rubber without turpentine. In the end, all he got from the diversion were a few days' reprieve for his muscles from tearing lump rubber to bits, and a few moments of comic relief to leaven his frustration.

A local boy named Jerry had been helping the inventor break down rubber for his experiments, and, inspired by Goodyear's enthusiasm, decided to try his own experiments. One night, he secretly pried open a barrel and dipped a pair of pants in the viscous liquid, then resealed the lid. After the trousers dried, Jerry arrived for work proudly wearing the pants. Goodyear was initially impressed. The trousers were smooth and pliable in the cool morning air. Then Jerry sat down next to a fire to start his chores. Within a few minutes, the warmth of the fire, combined with his body heat, fused the boy's legs to the pants and the pants to the chair. He wasn't harmed, except in pride, but the pants had to be cut from his body, much to the amusement of the residents of Congress

Avenue. When Goodyear conducted his own experiments with latex, he found the resulting products every bit as tacky as those made from the solid rubber-turpentine mixture. Turpentine was thus acquitted, and Goodyear was as far as ever from his goal.

Returning to his hard lumps, turpentine, and magnesia, Goodyear finally arrived at a formula that seemed to produce a reliably smooth finish—a half pound to one pound of magnesia per pound of rubber liquefied with turpentine. Though he made some items of solid rubber, most often he spread the mixture over canvas or other fabric. Piece after piece emerged smooth and hard. Elated, Goodyear excitedly relayed the news to Ralph Steele, and to every friend and acquaintance who would listen. Eager to repay Steele, and with visions of lifting his family out of Sodom Hill to the sort of comfort they had known briefly in Philadelphia, Goodyear hurried along with plans to manufacture rubber shoes. Of course, Goodyear was not the first man to rush into production before being certain of the quality of his products. That description effectively sums up the entire early rubber industry. Still, Goodyear's haste would cost him dearly, in time, money, and reputation, none of which he had to spare. Through an entire winter and into spring, Goodyear coated large pieces of canvas with his solution, then, with Clarissa's and Ellen's help, he cut and sewed the pieces into crude shoes. By early spring, several hundred pair were stacked neatly in the little cottage. Only then did Goodyear stop long enough to determine whether the shoes would be sufficiently durable to sell. He set them aside for several weeks, just as the Connecticut spring flirted with summer. After a few hot days, Goodyear returned to his wares, and found that they had congealed into a misshapen mess reminiscent of the life preservers he had first seen in the New York showroom of the Roxbury India Rubber Company. It was a crushing blow after a hard winter's work.

After the magnesia fiasco, Ralph Steele lost interest in rubber and drifted out of Goodyear's life. Sapped of his chief money source, Goodyear fell behind on his rent and soon ran out of furniture to pawn. Eventually Justice of the Peace John W. Payne delivered an eviction notice, in ponderous, Old English lettering, giving the Goodyears three days to gather their belongings and clear out. The date on the surviving eviction notice, January 27, 1835, is problematic, since the Goodyears were still in Philadelphia in early 1835. One possible explanation is that the date on the document is simply in error. The eviction form was preprinted with the year, so in January 1836 the justice of the peace may simply have used a form from the recently ended year. But whatever the explanation for the discrepancy, one fact remains clear: The Goodyears were now too poor for Sodom Hill.

Goodyear moved his family to a cheaper cottage in the country northeast of New Haven. Goodyear's reputation may have preceded him, for his new landlord demanded he put up security, in advance, against unpaid rent. The family presented as collateral a quantity of linen hand spun by Clarissa. Within a few months, with Goodyear once again behind on rent, Clarissa's linens, which should have been a family heirloom, went to auction. About the time of the move to the new home, William, still a toddler, died. Charles Jr., now three, became seriously ill as well but pulled through. Clarissa had given birth to six children and watched three die in infancy. As for Goodyear, he was thirty-six years old, well beyond the age of youthful fancies. Wracked with grief over the death of his children, and with guilt over his inability to provide for his family, and dogged by his own chronic ill health, he plodded on.

His reputation in Connecticut on the rocks, Goodyear looked just south to New York City, where an acquaintance named J. W. Sexton, who admired his perseverance, agreed to provide him a small room on Gold Street where he could live and conduct experiments. More important, the proprietors of a Manhattan drugstore, Silas Carle & Nephew, offered to provide the chemicals he would need. He was still using magnesia, but had begun experimenting with lime and white lead as well.

About this time William DeForest, his brother-in-law and former tutor, stopped at Gold Street to visit him. DeForest was now a prosperous woolen manufacturer in Naugatuck. Tall and athletic, he bounced up the three flights of stairs and knocked on the door. He was prepared for Goodyear to be living in poor conditions. Still, when Goodyear opened the door, DeForest was shocked. There was barely room for a bed amid the jumble of kettles, blocks of rubber, and chemicals. DeForest had never known Goodyear to be in good health, but now he appeared particularly gaunt, drawn, and pale. His hair was matted and his clothes were unwashed and stained with every manner of chemical. Goodyear smiled and moved toward DeForest but did not extend his hand.

"I don't know how to rub India rubber off," Goodyear said sheepishly, looking at his hands. "There's only one way—by rubbing more on."

DeForest smiled at Goodyear's feeble joke. He had not come simply to say hello, but to discourage Goodyear from further experiments and to convince him to return to his family and put his labors toward something productive. DeForest reminded Goodyear of the burden on Clarissa, and of the family's ever tenuous circumstances.

Goodyear gestured with his arms and said, "William, here is something that will pay all of our debts and make us comfortable."

DeForest glanced about the messy room, sighed, and said, "The India rubber business is below par."

"Well," said Goodyear, smiling, "I am the man to bring it back again."

Goodyear believed lime might reinforce the smoothing effect of magnesia on rubber. He spread his magnesia-rubber on strips of cloth, allowed them to dry, then dipped them in kettles of boiling quicklime and water. The smell must have been terrible, but for a time he believed himself to be nearing another breakthrough. The articles came out even harder and less prone to tackiness than did the magnesia mixture by itself. Once again, Goodyear announced precipitously to all who would listen that he had solved the rubber dilemma. He submitted sample products to the Mechanics Institute of New York, which was sufficiently impressed to award him a silver medal. With the new attention and apparent success, Goodyear was able to sell some rubber-coated draperies and other goods. He moved out of the cramped Gold Street room and into attic quarters at a boardinghouse called Holt's Hotel. His daughter Ellen came down from Connecticut to join him for a time. But rubber was not giving up its secrets so easily. The lime-dipping method soon revealed its own serious defects. Objects coated in lime proved to be susceptible to anything acidic. Even something as mild as apple juice washed away the hardened surface and caused the rubber to become as tacky as ever. Undaunted, Goodyear theorized that perhaps the lime coating was insufficient—perhaps the goods needed to be infused with lime during the mixing process, rather than simply dipped in lime water after the fact. Lime was too caustic for him to knead it in with his hands. Goodyear soon located a horse-drawn mill in Greenwich Village, operated by an obliging soul named Pike, who allowed Goodyear to mix the rubber there. The mill was three miles from Goodyear's quarters. Each morning, Goodyear set out with a gallon jug of slaked lime over his shoulder, and trudged the three miles on foot.

Goodyear's conception of rubber was always romantic rather than commercial. In addition to making rubber tougher, he sought to make it beautiful as well. A fanciful experiment led to an important discovery one day, after he decorated a piece of rubberized fabric with bronze. When the bronze coated poorly, Goodyear tried to salvage the piece by washing the bronze off with nitric acid. But nitric acid only further discolored the rubber, and Goodyear cast it aside. Several days later he happened across the piece. It was smooth and hard—a better finish than he'd ever achieved with magnesia or lime.

Goodyear set to work in earnest exploring this new clue, coating every piece of rubber fabric he made with nitric acid. He spent several months treating rubber in this way. He called it tanning, since it reminded him of tanning leather, and referred to the overall process as the acid-gas process. Each new breakthrough seemed to involve a different and potentially deadly chemical. Given Goodyear's frail constitution, the harsh chemicals, the small unventilated rooms where he worked, and the fact that he did much of his mixing by hand, it's a wonder that Goodyear lived as long as he did. Believing himself to be on the verge of a tremendous discovery, he conducted much of his work in closed rooms, lest prying eyes discover his secrets. One day, while mixing a nitric-acid solution, a cloud of the gas billowed from the mixing bowl, enveloping Goodyear. As he breathed in, he reared back in pain, his throat and lungs seared. He doubled over, clutching at his throat and coughing uncontrollably.

Goodyear spent six weeks in bed, much of the time in a delirious fever. When he recovered, he returned to nitric acid. Once again he found an investor. William Ballard was a prominent New York merchant, whom Goodyear had met at the Mechanics Institute. Ballard, excited by Goodyear's nitric-acid process and not dissuaded by (or unaware of) the failure with magnesia and lime, agreed to finance further experiments. Ballard paid to rent space in a factory equipped with steam power on Bank Street in Manhattan. Now that he had a backer, Goodyear rushed to Washington and applied for a patent, which was issued in 1837. As he always did when it came to rubber, Goodyear tirelessly promoted his new method. While in Washington, he managed to submit to President Andrew Jackson several samples of his work, including a print he had made on a sheet of rubber and rubber-coated bandages for military use.

Jackson sent a cordial reply on March 4, 1837:

> *I thank you for these samples of your skill in the new art in which you are engaged, and which I have no doubt will be found useful in a great variety of ways. I*

can only wish you success in the prosecution of your useful labors, and assure you that the sentiments of kindness which you express are cordially received and reciprocated by your humble servant,

Andrew Jackson

Goodyear received a similarly encouraging note from Senators Henry Clay and John Calhoun, after giving them samples as well. Back in New York, Goodyear started turning out hats, caps, aprons, and life preservers. The goods sold modestly to a public understandably skeptical of anything made of rubber. Still, it was a time of optimism for Goodyear, especially backed by Ballard's enthusiastic support. Ballard and Goodyear began looking for a larger space and located an abandoned rubber factory on Staten Island, one of the many that had failed after the first wave of rubber madness subsided. Goodyear sent for Clarissa and the children to join him. The family turned out aprons, wagon covers, caps, and so forth, all of them treated with Goodyear's patented acid-gas process. These products would eventually deteriorate in hot weather—the acid-gas process was yet another masking of rubber's flaws rather than a cure. His best mask yet, to be sure, but a mask nonetheless.

Ballard had invested several thousand dollars in Goodyear and the rubber business. Then a financial panic swept the country in 1837, and Ballard was essentially wiped out. Despite the promising products he was developing, Goodyear could not generate enough sales to continue long without Ballard's support. He returned to his old standby, the pawnshop, hawking whatever furniture the family had managed to accumulate. Goodyear invited his younger brother, Robert, and his family to Staten Island. Robert was welcome in no small measure because he owned some furniture, and also as an extra source of free labor.

Between the financial panic and rubber's ever more dismal reputation, Goodyear could not induce a single potential investor to so much as visit the factory and examine his goods. The lone consolation was that nobody much cared what became of the old factory and the equipment inside; there were no impatient landlords pestering the impoverished collection of Goodyears occupying the small cottage and factory. Goodyear now found himself not simply needing materials for experiments and production, but food for his family. Robert helped by fishing for dinner in the nearby bay.

Charles journeyed to New York one day in search of funds and happened across a creditor to whom he still owed money. Steeling himself for a serious rebuke, Goodyear was shocked when the man greeted him with a warm smile.

"Is there anything you need?" the man asked.

Detecting no sarcasm in the man's voice, Goodyear cautiously said that fifteen dollars would relieve some pressing needs. The man promptly produced the fifteen dollars, and sent Goodyear on his way with good wishes. This act of kindness so moved Goodyear that he recalled it in detail more than fifteen years later, though omitting the name of the benefactor. The act not only buoyed Goodyear's spirits, it reinforced his conviction that when his need grew most pressing, God would see to it that he was provided for.

Finally, he found another steady investor, a New York acquaintance named William Ely, who agreed to pay for production and manufacturing. They signed an agreement to split any profits from goods to be manufactured on Staten Island. But it soon became apparent that the Staten Island business was going nowhere. With business at a dead end, Goodyear turned his attention to Boston, where the rubber industry still had a pulse, however faint.

Goodyear had recently met John Haskins in New York. Haskins was one of the original founders of the Roxbury India Rubber Company in Boston, the first and best-funded of the rubber companies that had started in the early 1830s, and one of the few still in business (though on a greatly reduced scale) in the late 1830s. Haskins, at thirty-five two years Goodyear's junior, had been warm and friendly toward the inventor. Goodyear told Haskins how his own obsession with rubber had started in the New York office of the Roxbury Company. He told of some of his past trials and errors, and then showed Haskins some samples of sheet rubber treated with the acid-gas process. Haskins was impressed. The samples were a considerable improvement over the goods emerging from his own factory. He invited Goodyear to visit the Roxbury plant that fall. Without covering specifics, the two discussed the possibility of a business relationship.

The Roxbury Company had managed to stay in business where others had failed, in part because of its superior initial funding, but mainly because of the inventiveness of Haskins's partner and foreman, Edwin Chaffee.

To coat cloth, Chaffee would first mix rubber with the universal solvent, turpentine. He then added lampblack, which gave the mixture a dark, uniform color, and seemed to offer some protection from deterioration caused by the sun. The liquid mixture would be poured into a long, rectangular box with a narrow open seam at the bottom. The rubber flowed out of the seam in an even line onto cloth as it unwound slowly from a roll. The coated cloth was suspended on wires for about two days, depending on the weather, to allow the turpentine to evaporate and the rubber to dry. The process was slow and costly. Like Goodyear, Chaffee hated turpentine. The company went through thirty-six barrels of the stuff per week, a huge expense for a material whose ultimate function was to disappear into the atmosphere. Chaffee began working on an alternative to solvents. He favored a combination of friction and steam heat. Chaffee spent three years designing and building a thirty-ton calender, an immense machine for the time, that quickly earned the nickname the Monster. The Monster consisted of three large, hollow metal rollers, stacked one upon another and heated from the inside by steam. Workers inserted rubber between the rollers, which moved at unequal speeds, causing a slipping action that tore and further softened the material. The rubber would quickly become soft and pliable, capable of being mixed with additives. When the compound was ready, cloth would be inserted and automatically coated.

Chaffee's machine worked beautifully, saving the company time and money while greatly increasing production capacity. Unfortunately, the Monster's birth in 1837 coincided with the rubber industry's demise. In time, Goodyear's vulcanization discovery would revive not just the rubber industry but Chaffee's invention, which is remarkably similar to calenders still in use today. For now, though, Chaffee had by all appearances invented a marvelous machine whose time had already passed. One man who understood the significance of the machine from the beginning was Charles Goodyear. As soon as he heard Haskins's descriptions, he knew he had to go to Boston.

It's not clear how much money Goodyear's new partner, William Ely, invested in Goodyear's experiments, despite the existence of the seemingly formal business agreement. Goodyear does not refer to Ely by name in his memoirs, saying only that he "succeeded in obtaining a small loan from a friend" to travel to Roxbury to meet with Chaffee and Samuel T. Armstrong, an agent for the company. Still, it was Ely's cash that enabled Goodyear to make a journey that would provide a crucial turning point in his quest.

In September, Goodyear boarded a steamship for Boston, leaving Clarissa behind in Staten Island with the children in a bleak little cottage next to the defunct factory, far away from all that was familiar to her. He left her just fifty cents to cover family expenses. He wrote to Ely on September 21, 1837, describing in some detail a meeting with Chaffee and Armstrong. He closed the letter fretting that Clarissa might be hurt to know that he had written to his business partner before her: "Don't tell Mrs. G. I write you first, not that I love her, she knows it. More anon. Adieu."

Goodyear's casual, day-to-day approach to finances is revealed in an October 14 letter to Ely in which he happily reported, "I can go very much upon the credit system for board here." This arrangement allowed him to send ten dollars to Clarissa, the first money he had given her since leaving her nearly a month earlier. "I am so conscience smitten on this score that I shall keep a fund in reserve for this purpose in preference to anything else," he vowed.

Ely sent Goodyear $140 for expenses in Roxbury. A letter from Goodyear on October 18 offers a rare if cryptic glimpse at Clarissa's emotional state, and at Goodyear's feelings of guilt. Being stranded with no money seems to have driven even this most tolerant person near the breaking point. The only wonder is that it hadn't happened sooner. "Your favor of $140 is at hand," he wrote to Ely. "I [may] return a part of it with a request that you will accompany my wife and children to Roxbury. I can't say which of you I want to see most. I fear I carry the joke too far with Mrs. G. Physical evils she can endure like a heroine, and they do not harm but, unlike ourselves, if she encounters other trials, she is in no way qualified to endure them." Goodyear wanted Ely to accompany his family in part because he felt he could not fetch them himself, for fear of being arrested in Staten Island because of some old debt: "Mrs. G. must leave Staten Island, and who can shield me from the charge of absconded debtor? It is impossible for me to think of returning until I have accomplished my object, and she cannot remain there until then."

Whether the object in question was financial solvency or the perfection of his nitric-acid process is not known, nor is it clear just what arrangements were finally made for bringing the family north. In any event, Clarissa and the children moved to Roxbury in the winter of 1837–38.

The crew at the Roxbury plant welcomed Goodyear with enthusiasm. The industry's swift decline had chased away all quick-buck schemers, trend followers, and daydreamers. Those who remained adopted a sort of bunker mentality. Anybody still showing serious enthusiasm for rubber was a good friend. Chaffee's Monster machine had enabled the company to linger on life support, but there was no denying that the future of rubber looked grim. Goodyear put it well: "The public mind was completely paralyzed and disgusted with the subject, and it was of importance to the credit of any man in business, that he should not be known to have anything to do with it [rubber]; and much more to his credit, if he could show that he had never engaged in the speculation."

The precise details of Goodyear's financial arrangement with the Roxbury partners are not clear, but the deal was in any case a generous one for Goodyear. He was permitted space in the factory to conduct his tests, and was given access to the Monster and all the raw materials he needed on credit. He continued to revise his acid-gas process. He also developed an improved method for manufacturing rubber shoes. He sold the patent to a manufacturer in Providence, Rhode Island. John Haskins and Luke Baldwin, meanwhile, purchased from Goodyear the right to make certain goods under Goodyear's nitric-acid patent. They set up a separate shop in Lynn, Massachusetts, a few miles away.

The income from the patent sale and licensing Haskins and Baldwin gave Goodyear a temporary and unaccustomed infusion of cash. His family settled into rooms in a large old mansion-turned-hotel called the Norfolk House. Built in 1781 as the home for a wealthy lawyer named Jack Ruggles, the house had been converted in 1825 into a hotel, with the addition of a large hall for public gatherings. The Norfolk House was a center of local activity, its lobby always full of visitors. For Clarissa, the hotel represented major improvement over the bleak outpost at the abandoned Staten Island factory, and for a time the family lived in relative contentment. It was another false spring, but they enjoyed it while they could. Goodyear's aging parents, Amasa and Cynthia, came to live with them for a time.

Regular access to Chaffee's Monster freed Goodyear from dealing with turpentine. The machine also was capable of producing the most uniform sheets of rubber and rubber-covered cloth that he had ever seen. He experimented with various pigments and paints to color the things he made. He converted sheets of rubber into cream-colored piano and table covers with silver or gold borders, and ladies' capes with ornate designs. His imagination for fanciful new objects of rubber seemed limitless. And yet, to the growing consternation of his friends at the Roxbury India Rubber Company, what never seemed to emerge from all of Goodyear's creative genius were products that might turn a profit—and turn the rubber industry around. He would finish one fabulous piano cover or drape only to lose interest and move on to something else. Haskins's generous provision allowing Goodyear to buy materials on credit was like declaring an open bar to an alcoholic. When his debt to the company reached more than two thousand dollars, Samuel T. Armstrong, the agent, became fed up with Goodyear's endless puttering. In the spring of 1838 he ordered Goodyear to quit inventing and start manufacturing something, or to leave. Haskins, although always an admirer of Goodyear's, had little choice but to concur; the Roxbury India Rubber Company could ill afford Goodyear's extravagances.

Instead of turning his ready mind to more practical pursuits, Goodyear did what came naturally to him—he began looking for another town, another factory, another associate. His search led him to Woburn, Massachusetts, about a dozen miles north of Roxbury, and to the offices of an all but defunct enterprise called the Eagle India Rubber Company.

4

HANCOCK AND THE BRITISH MACHINE

If Thomas Hancock of London paid any attention at all to the hapless, sputtering American rubber industry during the 1830s, it only reinforced his own sense of his accomplishments in England. He was unaware of America's half-mad rubber messiah, Charles Goodyear, still laboring in obscurity. But even if Hancock had known about Goodyear, little in the American's litany of trials and errors would have given reason for alarm. By the time Goodyear soiled Clarissa's rolling pin with rubber in a Philadelphia debtors' prison in 1834, Hancock had been working with rubber for more than fifteen years, and had six British patents to his name. For nearly a decade, since 1825, Hancock had been a partner in and, increasingly, the guiding force behind Charles Macintosh & Company, the lone successful rubber company on either side of the Atlantic.

In addition to their country's vastly more developed manufacturing base, British rubber makers had a crucial natural advantage over the Americans—the weather. The sizzling summers and frigid winters of New England, where most American manufacturing took place, might have been designed specifically to expose rubber's great weaknesses. The British climate, like its people, was more reserved, less given to extremes, and hence provided a more benign environment for rubber.

Even in Great Britain, to be sure, rubber remained a weak substance, susceptible to stress and friction, and useful only for a limited number of products. But rubber hats, coats, and other products, if well made from good rubber, provided reasonable protection from the rain. Charles Macintosh & Company, under the guiding hand of Thomas Hancock, had actually been able to squeeze profits from the rubber business.

Hancock, unusual among creative people, was a consummate businessman. He was finicky, organized, obsessed with details, and ever mindful of profits and losses. In short, he was nothing like Charles Goodyear. One would be hard-pressed to invent two personalities more opposite. Their only notable similarity was their mutual obsession with rubber. If you needed to support your family, and had the choice of working for Hancock or Goodyear, you would choose the former without hesitation. Hancock built factories and made money. Unlike the mercurial Goodyear, always chasing some new and more elusive dream, Hancock never entered any venture without weighing the cost against the potential market. When Hancock insured his London factory with the Guardian Assurance Company for £1,100 (paying an annual premium of £8.9), he was able to personally itemize every asset, including a weighing machine worth £4, a gas meter worth £5.17, and lathe tools and screw plates at £18. Even if Goodyear could have untangled his assets from his debts (which he could not), no insurance underwriter in his right mind would have issued a policy to a man who was a walking monument to risk.

Hancock enjoyed robust good health throughout his life, whereas Goodyear's medical history describes a sixty-year process of dying. Goodyear was clumsy around machines; Hancock was a mechanical wizard who designed and built much of his machinery himself. In experiments, Goodyear was spontaneous, impatient, and romantic. He could hardly wait for the results of one test before his mind wandered to something still more promising—a quality that drove his business partners to distraction. Hancock was meticulous and focused to a fault. One experiment took him thirty years to complete. His object was to test whether rubber was truly waterproof, or might in fact allow some infinitesimal amount of moisture to seep through. On a crisp fall day in 1826, Hancock filled a heavy bag with water. The bag was made of a thin sheet of rubber with an exterior of fustian, a coarse, sturdy cloth made of cotton and flax. He sealed the bag hermetically and hung it in his laboratory. For the next thirty years, Hancock retrieved the bag every few months, weighed it, and carefully recorded the measurements in a log book. The bag, which had weighed a little over a pound on that first day, dropped to an even pound by October of 1835, just over thirteen ounces by 1849, and under four ounces by May of 1854. In 1856, Hancock raised his seventy-year-old hands to retrieve for the final time a bag that forty-year-old hands had first hung. He recorded his impressions excitedly at the time: "I have just now, 1856, cut it open; it is quite dry, and weighs three ounces twelve drachms, proving that rubber is not absolutely impermeable to water, but admits of a slow and gradual absorption of moisture through its substance." As Hancock conceded, somewhat sheepishly, "This slow evaporation does not interfere with [rubber's] efficiency for ordinary purposes." The experiment said as much about Thomas Hancock as it did about rubber.

Hancock was born in 1786, fourteen years before Goodyear, in Marlborough, in rural Wiltshire County, a hundred miles west of London. The spire of the St. Peter's Church tower loomed over the town's central marketplace, which attracted farmers and craftsmen from around the county. Hancock was the second of twelve children, eight of them boys, born to Elizabeth and James Hancock, a cabinet maker and timber merchant. Hancock attended a sturdy, brick school in Marlborough, then started his career in various "mechanical pursuits." By 1815 he had moved to London, his principal home for the rest of his life.

Thomas Hancock inherited from his father the craftsman's touch, the hands of a builder. He was no theoretician; he gravitated toward inventing from the practical perspective of a professional mechanic and artisan. The time he spent in his father's shop taught him a love and feel for wood. He developed a facility not just for building things, but for fashioning his own tools. This would offer a crucial advantage when he began to experiment with rubber, for which there were few if any ready-made tools. He started in business in London as a coach maker, a natural extension of his woodworking background. His first partner was his younger brother John. The Hancock brothers together were an extraordinarily inventive and productive lot. Six of the eight would turn to inventing, compiling between them forty-seven British patents—fourteen belonging to Thomas. Next to Thomas, Walter, thirteen years his junior, was the most productive. Walter Hancock started his career as an apprentice jeweler and watchmaker, then switched to engineering. He became infatuated with the idea of steam-powered carriages that would run on regular roads, to compete with horse-drawn stagecoaches. Although the idea ultimately failed with the advent of railroads, Walter Hancock built ten steam carriages that ran regular service connecting towns in various parts of England, and is now regarded as a pioneer in steam transportation.

After just a few years in the coach-making business, Thomas Hancock found himself immersed in his own passion—rubber. He recalled in his memoirs: "I have no very clear recollection when I first began to notice the peculiar qualities of India-rubber, but well remember that the more I thought about it and tested its properties, the more I became surprised that a substance possessing such peculiar qualities should have remained so long neglected, and that the only use of it should be that of rubbing out pencil marks."

He began his experiments in earnest around 1819. The rubber that early European manufacturers such as Hancock used in experiments and manufacturing arrived from South America in smoked, multilayered, irregular globs called biscuits. Rubber historian Austin Coates cites one colorful description by a nineteenth-century journalist, of "skinny shreds, fibrous balls, twisted concretions, cheeselike cakes, and irregular masses." Each layer of rubber differed materially from the others. One would be perfectly smoked and free from impurities, while the next would be scorched and flecked with bits of ash, dirt, or whatever else happened to be floating in the jungle air when the latex flowed.

Hancock wrestled with the material as best he could. Just as Charles Goodyear would a few years later, Hancock cut rubber into small chunks and mixed them with turpentine. But the solutions came out thin and drippy, and they dried unevenly. So he concentrated on small items that he could make directly from the imported biscuits. His first English patent, dated April 29, 1820, was for "articles of dress." Hancock cut long, thin strips of rubber from the biscuits and hired seamstresses to attach them to the wristbands of gloves, to stockings, and "to such parts of the apparel and dress of women as require to be kept close to the person." The result, at least in theory, was a snug fit that did not require cumbersome straps or ties. Yet even in this modest employment, rubber showed its weaknesses. The seamstresses had to poke needles through the strips to attach them to garments, but a small hole was enough to start a tear that enlarged each time a man slipped a glove over his hand, or a lady adjusted her rubber-laced undergarments. Customers returned piles of Hancock's goods with rubber strips that had simply snapped. Hancock minimized this problem by cutting strips thicker on the ends, where the holes went, but thin at the middle, which needed to stretch.

Cutting usable strips from odd-shaped biscuits was difficult, tedious work. The biscuits quickly came to resemble Christmas hams set at by a clumsy carver. Waste rubber piled up in irregular chunks. The thrifty Hancock, abjuring waste in any form, began in 1820 to look for a way to use these scraps. His solution, true to his training and nature, was to build a machine.

Hancock's plan was to make a sort of grinder that would tear and gnaw rubber chunks into pebble-sized bits that he could more easily mix with solvents. He built a prototype with two pieces of wood bolted together to form a block with a hollow, cylindrical center. The circle was studded with metal teeth protruding inward. Inside the hollow area was a smaller cylinder, with teeth pointing out. Hancock fed rubber chunks into the cavity through an opening at the top, then turned the crank, feeling it become stiffer and stiffer with each turn. When he opened the lid to see the results, he was at first surprised and then delighted by what he found inside. Instead of the pile of loose pebbles, Hancock found a single wad of rubber. He had not anticipated the effect of heat generated by the friction of constant tearing and shredding. The ripping motion made the rubber sticky, so the pieces massed together like soft, moist dough. Inadvertently, Hancock had discovered a way to manipulate rubber without using solvents. Now Hancock could create blocks of uniform size and, more important, uniform consistency. From these soft blocks he could cut pieces into any shape he wanted.

Thrilled, Hancock cranked furiously away until the wooden model literally came apart. He quickly commissioned Hague and Topham, London engineers, to build a cast-iron prototype according to his precise specifications. The machine could hold only a couple of ounces of rubber at a time, but it performed flawlessly. Today, that machine sits in a climate-controlled room in the London Science Museum's labyrinthine storage warehouse, on a shelf with other artifacts from the remarkable collection of early rubber, gutta-percha, and plastics. The size of a small coffee grinder, it's a marvelous piece of workmanship, fully justifying Hancock's reputation for mechanical genius. Heavy and solid, yet sleek and modern-looking, it belongs in appearance more to the early twentieth century than the early nineteenth. It has a removable hand crank of steel and a pair of interchangeable steel cylinders, one with large teeth for coarse tearing, the other grooved for finer work.

Hancock soon ordered a larger metal model capable of handling several pounds of rubber at a time, and, after that, machines that could hold several hundred pounds and that delivered solid blocks of rubber six feet long, a foot wide, and seven inches thick. Nobody else could turn out rubber of such fine and uniform size. As these fine pieces of rubber emerged from Hancock's shop, competitors and customers alike naturally began asking questions. Hancock chose not to patent his device, which would have necessitated making the details public. Instead, he kept the machine hidden from prying eyes. He required his few employees allowed to operate the machine to take vows of silence. To mislead the curious, Hancock labeled the machine a "pickle," to imply some sort of chemical soaking process. The deception worked; he held the secret for more than a decade. Only after the technology became known did Hancock publicly adopt a more accurate term, the masticator.

The masticator was the most significant breakthrough in rubber to that time, and remains one of the most important in the history of rubber. Hancock was never shy about trumpeting its importance. In his memoirs, published in 1857, eight years before his death, he wrote: "I wish here to remark that the discovery of this process was unquestionably the origin and commencement of the India-rubber manufacture, properly so-called: nothing that had been done before had amounted to a manufacture of this substance, but consisted merely in experimental attempts to dissolve it; and even this had never yet been effected for any useful purposes."

He was seventy years old when he wrote those words, to which he added the bitter complaint that his masticating process "has been from time to time adopted by others, and even introduced into specifications without the least acknowledgment" by people who "adopted it as if it were their own." This charge fairly dripped with irony, considering that by the time Hancock's memoirs appeared he had already done the same, and more, to Charles Goodyear. But all of that was still far in the future when the thirty-four-year-old Hancock first turned the crank on his masticator and produced the finest and most consistent rubber the world had seen.

Hancock immediately began searching for new products that would take full advantage of his discovery. Now he could cut rubber strips for gloves and undergarments from standard-sized blocks, improving quality and eliminating waste. He could also produce high-quality and evenly shaped pencil erasers, which sold well through stationers. But he quickly realized that the masticator was good for much more than enabling people to erase pencil marks while wearing snug-fitting undies. But what?

The answer, one that ultimately enabled Hancock to consolidate much of Britain's rubber industry under his own direction, came from a Glasgow chemist named Charles Macintosh. Macintosh never intended to get into the rubber business at all. He was contentedly manufacturing industrial chemicals in Scotland when he stumbled onto the process that would ensure his immortality. This happy accident hinged in turn on a series of unforeseen and barely related events set in motion when another eccentric genius found a way to bring light to a darkened world.

Until the turn of the nineteenth century, Great Britain's nights flickered dully in the smudgy, muted glow of whale-oil lamps, tallow, and candles. These implements were dirty, messy, and a terrible fire hazard. Travel at night was difficult or impossible, and even in large cities few ventured out on foot after dark. Oil street lamps cast such paltry light that pedestrians could scarcely distinguish walkways from potholes or other lethal impediments. But the danger of falling was nothing compared with the threat posed by pickpockets and muggers, who flourished in the ample cover of darkness. The well-heeled could hire a "link boy" to walk several paces ahead, holding a lantern to light the way. But even this luxury entailed risks—it was not unheard of for link boys to lead unsuspecting charges to some lonely street, only to rob them and disappear into the night. Indoors, life after dark was less harrowing but equally limited. Beyond reading, writing, and simple chores, work was out. Factories closed at dusk. Candles burned brighter than oil, but they were the product of skilled craftsmen. Their brief usable life made them prohibitively expensive for all but the wealthiest people, who might burn several hundred dollars' worth (and, sometimes, their homes) in the course of a party. By and large, human activity remained prisoner to sunset.

Then a shy, Scottish-born foundry engineer named William Murdoch began toying with the idea of lighting with coal gas—produced when coal is compressed and heated in an airtight chamber called a retort. His neighbors in Redruth, Cornwall, had always considered Murdoch an odd bird. He kept to himself and some said he wore hats made of wood. They were convinced the man had lost his mind in 1794, when he laid pipes from his garden retort into his home. But the experiment filled his house with a most wonderful light. Intrigued, his company allowed him to rig a lighting system for its Soho foundry. The gas burned clean, the glow was exceptionally bright, and the cost was a fraction of that for oil or candles. For eyes trained in the sorrowful murk of oil lamps, gaslight was like stepping from a cave into a brilliant day. Factories and theaters were the first to embrace the new technology. By 1815, London alone had four thousand gaslights. By 1819, a network of 288 miles of pipe supplied more than fifty thousand burners throughout the city. Within a generation, over the protests of the whale-oil industry and doomsayers who insisted Murdoch had harnessed the devil, the entire kingdom was converting.

The sudden demand for gas generated prodigious amounts of thick, noxious sludge called coal tar. This was not the age of environmentalism. The London gasworks tossed its waste into the Thames, while Edinburgh befouled the Firth of Forth. In Glasgow, coal tar found its way into less scenic spots—mainly, deserted coal pits and quarries. But these were quickly filling up.

In 1819, Charles Macintosh offered to buy Glasgow's entire output of coal tar. Born in 1766, Macintosh was the son of a prosperous chemical manufacturer. Seldom do birth and predisposition agree so happily, for Charles Macintosh was a born chemist. When his father steered him into apprenticeship in a counting house, Macintosh chafed and spent every free moment attending scientific lectures in Glasgow and Edinburgh. When he came of age, he inherited his father's works and embarked on a career that would outshine the old man's by a wide mark. Within a few years he was doing a brisk business in sal ammoniac, sugar of lead, acetate of alumina, and other chemicals widely used in British manufacturing of the time. Successful as Macintosh was, his name would surely have gone down in obscurity had he not decided to buy coal tar from the Glasgow gasworks, which agreed at once to unload every cursed ton for next to nothing.

Macintosh wanted the tar for its ammonia, which he needed to make cudbear, a purple dye used in wool and silk. His employees did not protest the arrival of this vile coal sludge. Vile sludge was a step up. For years they had obtained their ammonia from the collective urine production of the human population of Glasgow. Lord knows what offenses one had to commit in a previous life to wind up consigned in this one to a career on the Macintosh pee patrol. Each day, workers harvested up to three thousand gallons of their fellow townsmen's finest, from buckets and casks strategically placed around the city (pubs, no doubt, providing a particularly rich vein). Wagons full, they sloshed their way back to the works, where they distilled ammonia in a factory that must have smelled perpetually like the men's room of Hell's saloon on dollar-a-pitcher night.

Collecting urine was as inefficient as it was unpleasant; thus coal tar, with its ampler and more reliable ammonia content, put an end to one of the weirder chapters in manufacturing history. Yet the new process left Macintosh with another problem, one that brings us back, at last, to the story of rubber. Robbed of its ammonia, coal tar left behind a harsh, volatile liquid called naphtha, for which there was little known use. Macintosh could hardly return the naphtha to the gasworks, but where could he dump it? Then he read a speculative article in a scientific journal suggesting that naphtha might be used to dissolve rubber, which had lately begun arriving in ever greater quantities from the Amazon. Experimenters had already established that turpentine or ether would dissolve the stuff. But turpentine was expensive and did not evaporate well. Ether was too volatile and exotic to be of much practical use. Macintosh's experiments soon showed that naphtha evaporated quickly and more thoroughly from rubber than did turpentine, leaving molded or spread rubber in more or less its original condition. It had the additional benefits of being dirt cheap and plentiful.

Rubber's mysterious qualities seduced Macintosh as they had Hancock. He experimented with spreading his liquefied rubber-naphtha mixture over cloth. The emerging fabric was still too tacky to be exposed directly to the skin or the elements. Then an idea, as simple as it was brilliant, struck him. He laid out a stretch of fabric and coated it with a layer of rubber and naphtha. Before the naphtha evaporated, he laid on another piece of fabric. The sticky interior layer cemented the fabric layers together. The result had the agreeable exterior of cloth, but with a layer of waterproofing to keep out the rain. Macintosh took out British patent number 4804 to protect the process. He and three partners, brothers named Birley, set up a separate company in Manchester, England, and began manufacturing waterproof canvas for ships, clothing for sailors, and, famously, raincoats. To be sure, Macintosh had not eliminated rubber's great flaws, merely subdued them. Purchasers of Macintosh coats were warned on the label not to stand too near a fire after taking refuge from the British rain, lest they weld their coats to their backsides. Still, Macintosh's material was well-suited to Britain's wet but relatively temperate climate, and his garments were among the first, and certainly the most celebrated, mass-produced items made from rubber. He gave his raincoats the cumbersome name "water proof double textures." But buyers quickly rejected that in favor of the name that survives to this day, with a slight spelling change—the mackintosh.

Even with the advantage of Macintosh's innovation, the Manchester factory struggled at first for survival. Goods came out in dreadfully uneven quality. This had much to do with the uneven quality of the rubber going into them. But there were other problems. Macintosh needed large amounts of naphtha to dissolve his rubber, which meant that his goods dried painfully slowly, and some refused to dry at all. On top of that, naphtha's harsh smell was even worse than rubber's. Because of the high proportion of naphtha to rubber, naphtha seeped into the fabric before the mixture could dry, resulting in a pungent and persistent smell that was more than some consumers could bear.

In 1825, Hancock, now thirty-nine years old, purchased a license from Macintosh to manufacture raincoats and other goods using the double-texture process. To reduce odors, Hancock cut his naphtha with the purest turpentine he could find. More important, rubber mixed and softened in Hancock's masticator needed only half as much solvent of any sort in order to be spread onto cloth. Hancock's garments, therefore, dried more reliably, and smelled better than did Macintosh's. Shortly after taking the license, Hancock offered to supply his own mixture to Macintosh's company. Macintosh, annoyed, brushed off the overture.

Hancock's was a small-time outfit compared with Macintosh's extensive chemical works. Who was Hancock to think he could supply rubber to the man who had invented double-textured cloth? But annoyance turned to alarm as Hancock's superior cloth began outselling Macintosh's.

Macintosh & Company hastily struck a deal with Hancock. Instead of being merely a licensee, Hancock would now manufacture rubberized cloth directly for Macintosh. Hancock, sensing his new position of strength, demanded the right to stamp his own name on all products made from his cloth, and that Macintosh not sell anything to directly compete with Hancock's products. Moreover, Hancock would not have to reveal the secrets of his masticator, which still went by the name of "the pickle."

Within a few years Hancock became a full partner in Charles Macintosh & Company. Although the firm continued to bear Macintosh's name, Hancock steadily assumed practical control, often ordering design changes that improved the reliability and salability of products. A wearer of an early Macintosh raincoat might walk confidently out into a London drizzle, only to finish his walk, remove his coat, and find himself soaked. The problem was that Macintosh had contracted with independent tailors, who used the same stitching methods for raincoats as for conventional garments. Rain seeped through the exterior needle holes, collected, then coursed along rivulets at the interior seams. This was a particularly invidious way to become wet, and the defect all but negated the waterproof properties of the rubber. Hancock urged the tailors to use larger sections of rubberized cloth to minimize cutting and stitching, and to stitch only the inner piece of fabric, leaving the rubber and outer cloth layer without holes. When the tailors resisted changing their accustomed ways, Hancock hired other tailors to work in the Macintosh factory under the eyes of supervisors.

Hancock also changed the style of the coats. Tight coats may have been fashionable, but rubber didn't just keep rainwater out—it kept perspiration in. The buildup of sweat and body heat was unpleasant for the wearer and harmful to the coats, when the rubber layer became soft and sticky. Hancock directed the company to make only loose-fitting cloaks with sufficient room underneath for air circulation. With these changes, comfort and reliability rose, and so did sales. Before long, officers of the Guards were wearing light drab cloaks of cambric cloth to field exercises. The Duke of York could be seen reviewing his soldiers in a splendid cloak of blue cloth lined with crimson silk, with a snug filling of Thomas Hancock's finest masticated rubber between the layers.

5

WOBURN

Hancock, proud of his working-class origins, always insisted he was a mechanic, not a scientist. Still, he played a significant role in shaping modern understanding of rubber. He was, for example, among the first to notice and study the deleterious effect of sunlight on rubber. And it was Hancock who supplied Michael Faraday, the most eminent scientist of the period, with a rare sample of pure liquid latex, which Faraday used to give the first accurate molecular recipe for rubber—90 parts of carbon to 9.11 of hydrogen. Hancock had ordered the sample carefully preserved during the long journey from the Amazon in hollowed-out canes with rubber stoppers at the ends. On the mechanical side, Hancock in 1836 installed in his London shop the first steam engine to be used in rubber manufacture. Nicknamed the Grasshopper, the engine ran almost continuously for more than eighty years before finally being dismantled and removed in 1922.

Yet for all of his contributions, Hancock for most of his career remained curiously inactive in the one area that mattered most—trying to find a way to make rubber impervious to heat or cold. Hancock was shrewd enough to know that the solution to this problem represented the Holy Grail of his profession. Yes, he had turned rubber into a successful enterprise where many others had failed. But even the best-run business could only go so far based on an inherently flawed commodity. Rubber as yet had little if any reliable place in machinery. Extended friction, pulling, or compression generated too much internal heat. The stuff tore too easily. Solvents reduced it to liquid. A substance so eager to fail under pressure could never carry much of an industrial burden. A fire hose that cracked when the weather dipped below freezing was all but worthless. A life preserver that deteriorated in the hot sun was no life preserver at all. Oh, but the possibilities if the problem could be licked!

In fact, Hancock did work on the problem for a time early in his career, but then he gave up. His memoirs offer little about the specific nature of these experiments, but the results clearly disappointed and frustrated him, and the "rather severe loss" entailed in fruitless trials unnerved him. Every additive, every recipe, every new mixing method he tried consumed precious quantities of rubber, not to mention time he could have devoted to actual products. In the end, financial success proved to be his greatest barrier to discovery. It made him cautious. Hancock the responsible businessman simply would not give Hancock the inventor sufficient leash to tackle the problem in earnest. Faced with imminent failure, he might have gambled on the all-or-nothing challenge. But Hancock was making money for himself and for Charles Macintosh & Company.

To put his mind at ease, Hancock allowed himself to believe for years that the thing simply could not be done. If the father of the rubber industry could not solve nature's maddening riddle, nobody could. As Hancock consoled himself with this pessimistic, oddly comforting notion, Charles Goodyear, in his wild, fitful, but tenacious way was moving inexorably closer to proving him wrong.

On a midsummer's day in 1838, a farmer looking up from his field in Woburn, Massachusetts, might have noticed with mild curiosity a chaise carriage pulled by a single horse as it crested Wood's Hill and rolled down Railroad Avenue toward East Woburn. At the reins was a slender, unhealthy-looking man in his late thirties, with a shock of dark hair, high cheekbones, and deep-set eyes. Seated next to him was an older man, in his sixties, but similar in appearance to the younger man. After his falling out with the managers in Roxbury, Charles Goodyear needed a new place to continue his experiments. He had driven out to Woburn from Roxbury with his father, Amasa, the old inventor, in hopes of inspecting the factory of an ailing enterprise called the Eagle India Rubber Company.

Goodyear pulled the chaise to a stop next to the Eagle mill. The two-year-old mill was sixty feet long and thirty feet wide, two stories tall. Woburn investors had poured $30,000 into the structure, and the money showed in the sturdy construction and decorative touches. The basement was made of stone and the upper portions of wood. Inside were spruce floors and plastered walls. The north end of the building sported a cupola with a hanging bell. The building straddled a canal dug from the nearby Aberjona River. Water entered from the west through a raceway in the stone basement, where it turned a fifteen-foot wheel.

Goodyear helped Amasa down from the carriage, walked to the front door of the building, and knocked. The Goodyears had come to see Nathaniel Hayward, the superintendent. Goodyear had heard of the distress of the Eagle plant and hoped Hayward might consider selling out. The man who politely but cautiously answered the door was stocky, with a pleasant, rounded face, a thick neck, large eyes, and wavy black hair receding on top but covering half of his ears and falling long in the back. Even more than Goodyear and Hancock, Nathaniel Hayward was the model of the homespun inventor, moving forward through sweat and determination rather than scientific training. Goodyear and Hancock, at least, had a fair education for the time. Hayward was illiterate through much of his life, marking documents with an "X" into his thirties and only later in life learning to read and write.

Hayward was six years younger than Goodyear, born January 19, 1807, in Easton, Massachusetts. Through his early twenties, Hayward operated a carriage shop and livery in Boston. In the spring of 1834, Hayward bought a large piece of rubberized cloth from the Roxbury India Rubber Company, as a rain cover for a carryall carriage.

Returning to his shop, Hayward unrolled the piece and fitted it over the top of the carryall. He used the overrun to fashion side curtains, then stood back to admire his handiwork. The rubberized cloth was smooth and supple, and stretched snugly over the carriage ribs. When the first cool spring rains arrived over Boston, Hayward eagerly hitched up a horse and took the carriage for a test ride. To his delight, the coach stayed dry and cozy inside. Unlike leather or oilcloth covers, which would eventually become saturated, the rubber cover completely repelled the rain. It didn't even need to be wiped dry when he returned to the stable. The only objectionable quality was the pungent smell—a curiously deep vegetable odor, like the bottom of a mulch pile. But the odor didn't overly bother Hayward or his customers—what was one more strong smell in a stable?

But rubber inevitably failed even its most ardent admirers. For Hayward the day of reckoning came when the hot, muggy summer of 1834 settled over Boston. He had used leather straps to tie back the rubber curtains on the sides of the carriage. When he went to untie the straps one day, the curtains, instead of unfolding, stayed bunched up, the creases fused together. Alarmed, he ran his fingers along the top of the carriage cover. The surface had been so smooth and cool a few weeks earlier. Now the rubber grabbed at his fingers as they passed over. The smell, once a minor inconvenience, was now revolting. In short, his beloved carriage cover was useless. Had Hayward been an unimaginative, incurious man, he most likely would have torn the cover and curtains from his carryall and marched down to the offices of the Roxbury India Rubber Company to demand a refund. Certainly, the proprietors would not have been surprised to see him. Angry customers were all too familiar that summer of 1834. The carriage cover and curtains would have taken their place in the ignominious pit behind the factory where the company buried its returned goods. Hayward might have gone back, grumbling, to his stable, refitted his carryall with an old leather or oil-cloth cover, and forgotten about rubber altogether.

But Nathaniel Hayward saw a mystery that needed solving. To some men, rubber's very failures acted as a challenge. And Hayward was tired of the drudgery of the carriage trade—"I was sick of the business I was in," he flatly recalled. He did not return the rubber carriage cover; instead, he started to experiment. As with Goodyear, his experiments were crude, self-designed, and wholly empirical. Working in his shop, Hayward began by mixing rubber with turpentine, then spreading the compound onto a small iron square and mixing in lampblack with the edge of a knife. He would then spread the mixture with his knife on strips of cloth two inches wide by six or eight inches long, and allow them to dry. Hayward worked on various mixtures through the summer and into the fall. As his interest in rubber grew, it became harder and harder to keep his carriages running. One thing had to go. Hayward stayed with rubber.

It was a bold decision for a man with a wife and young children, to abandon a business that, as mundane as it might have been, at least guaranteed an income. Hayward sold his equipment and stable and returned home to Easton, where he rented an old, water-powered mill known as the Quaker Leonard Factory for $12.50 per month. One of Hayward's first innovations was a mill for mixing liquefied rubber with other ingredients. It consisted of a tub with arms projecting inward. Inside was a revolving shaft, also with arms protruding. There were holes in the bottom of the mill. A water-turned pulley rotated the shaft, mixing the ingredients against the arms. A wiper bar at the bottom of the tub forced the mixture through the holes in the bottom. The mixture would then be poured into the spreader machine, which consisted of three large drums, or rollers, through which cloth was fed. Hayward showed his early samples to an acquaintance named Edward Haynes, Jr., of Woburn. Haynes in turn showed several other potential investors. Each contributed $25 so Hayward could continue his experiments. Although the investment didn't result in any salable products immediately, the investors were impressed enough with the samples to form the Eagle India Rubber Company in 1836, and to name Hayward foreman.

Woburn had fewer than three thousand residents in the mid-1830s. Boston, the center of culture, learning, and international trade, lay twelve miles to the south. The textile colossus of Lowell, the nation's oldest and largest manufacturing city, lay fourteen miles to the north. Woburn, large enough to have ambitions but too small to compete with Boston or Lowell, struggled to find its place in the middle. Haynes and other Woburn residents were itching to lift their city from the shadows of their two imposing neighbors. When the Boston and Lowell Railroad sliced through the town in 1835, adding quick access to the outside world, Woburnites saw their chance. All they needed was a product.

The first idea was to cultivate oranges, a venture that failed for reasons that now seem painfully obvious. Next, a group of investors decided that Woburn's future lay in silk. They would raise fat, young silkworms on tender leaves from mulberry orchards, spin the silk into thread, and weave the thread into fine, finished silk garments that would roll out of the mill and off to the railroad depot. Various souls had been trying to establish a silk industry in North America since 1622, when King James I of England sent silkworm eggs and mulberry trees in hopes that the colonists would supply raw silk for royal manufacturers back home. It was difficult and labor-intensive work. A few colonies, particularly Georgia and Connecticut, manufactured silk, but it never caught on the way the king had hoped and the nascent industry died with the American Revolution. Then in the early nineteenth century, with thousands of would-be entrepreneurs searching for new industries, the nation suddenly rediscovered silk. An industry that had struck colonists as a royal burden now struck a nation of newly free, ambitious entrepreneurs as the next path to riches. The silk fever equaled the rubber fever in fervor and folly. Suddenly, everybody had to be in the silk business. In Massachusetts alone, companies popped up in Northampton, Dedham, Nantucket, Newburyport, Roxbury, and Framingham. Other plantations and factories started in Rhode Island, Connecticut, New Hampshire, New York, New Jersey, Pennsylvania, Ohio, Virginia, Kentucky, and Tennessee.

It was this fever that caught up the good burghers of Woburn. They decided that East Woburn, with its broad, flat plains and proximity to the railroad, which ran just on the other side of Walker's Pond, was perfect for the planting of thousands of young mulberry trees. The newly formed Woburn Agricultural & Manufacturing Company bought three hundred acres of the plains in 1835, and converted a grist mill into a projected silk factory. Power was to be supplied by a mile-and-a-quarter-long canal cut from the nearby Aberjona River.

But like most who latch onto a trend, investors in the Woburn Agricultural & Manufacturing Company were late to get in and early to lose out. They planted more than ten thousand young mulberry trees. Instead of the standard hardy white mulberry, they selected the rarer and more delicate Chinese mulberry, said to be more appetizing to the worms. They also planted sixteen pounds of mulberry seeds, anticipating a yield of nearly four and a half million trees. They predicted that they would begin manufacturing silk within a year. What the Woburnites lacked was any expertise in actually running a silk farm. The Chinese had been perfecting the art for nearly three thousand years, the Europeans for nearly a thousand. If confidence in their Yankee ingenuity allowed the Woburnites to overlook this minor detail, the first cold winter gave them a harsh lesson. A series of killing frosts in the first winter destroyed huge numbers of saplings before the first cocoons could be harvested. About that time, the speculative bubble popped on the silk industry in America as investors in other towns around the Northeast ran into similar problems. Chinese mulberry trees became almost worthless overnight. The Woburn Agricultural & Manufacturing Company, having paid five dollars per sapling just a few months before, sold off surviving saplings for a penny apiece. Without a thread of silk to show for their efforts, the managers auctioned off most of the silk farm property.

Their silk dreams shattered, the investors set out to rescue their fortunes through rubber. Although they would prove to be as woefully behind the trend in rubber as they had been in silk, their decision would secure a place in history for the town by drawing first Nathaniel Hayward, and then Charles Goodyear. The Eagle India Rubber Company opened with Woburnite Oliver B. Coolidge as president, and Samuel Sweetser, Jacob Richardson, Luke Bigelow, and Edward Haynes, Jr., as directors. Hayward, as foreman, was given a three-year contract at $1,000 per year. But most of that money was theoretical—Hayward himself invested $1,000 in the company, some $700 or $800 of which he paid off through labor. Luke Bigelow, superintendent, earned around $800. Several workers transferred from Easton; others were hired in Woburn. The men earned $1.00 a day; women (most of them teenage girls) received between $3.00 and $3.50 per week. So certain were the directors that they were onto something big that Hayward had to sign a confidentiality agreement stating that he "will not impart or give information to any person or persons whatever respecting the secrets or art of manufacturing India Rubber."

Secrecy was the byword of all manufacturing. An unscrupulous worker with a keen eye might walk off with production methods and start his own business, or sell to some competitor. Hayward would mix his compounds away from the main floor, in a small room in the south-east corner of the basement. The room was kept locked, even when directors visited the factory. Hayward whitewashed the windows to prevent spies from learning his secrets.

The factory produced mainly shoes and aprons, which were shipped to Boston and elsewhere for sale in shops. But the basic weakness that had ruined Hayward's first carryall cover dogged his own products. Hayward may have been a tenacious worker, but the rubber still melted in the heat and stiffened in the cold. Moreover, consistency proved to be a major hurdle. One batch of products would come out all right, while the next, using identical methods, would prove tacky from the start.

"The business was very uncertain and unprofitable on account of the defects in the articles made," Hayward would later recall. "They became clammy and sticky when exposed to heat, and were stiffened by the cold. The goods which I made were often turned back on my hands on account of these defects. Having no other means of obtaining a livelihood, I struggled along in the business . . . making shoes, life preservers, and some coats, working all the time with my own hands and occasionally hiring a man to assist me. The value of the goods I made was not worth more than two thousand dollars a year, and the profits not more than three or four hundred dollars a year. The difficulties which I encountered led me to see that the only hope of making goods business profitable was in overcoming the defects of the goods."

Before long, the Eagle India Rubber Company was in serious trouble. Hayward and William Humphrey, one of Eagle's employees, bought the company from the original investors. The price couldn't have been much—Hayward in essence bought the company with wages the company owed him, simply freeing them from his contract. The directors, still licking their wounds over the silk farm debacle, were only too happy to unload the failed rubber plant, if only to avoid sinking further into debt. Another employee, Timothy Newell, joined the venture briefly. Within nine months, both Newell and Humphrey dropped out. By mid-1838, Nathaniel Hayward was alone, the sole captain of what was by all appearances a sinking rubber ship.

It was about this time that Goodyear knocked on Hayward's door and introduced himself. Goodyear and Hayward exchanged pleasantries for a few minutes. Then Goodyear asked for a tour of the factory. Hayward declined. Goodyear asked politely if Hayward would consider selling the factory. Hayward said he would consider selling for the right price. The exchange was noncommittal. Over the next few weeks, Goodyear sent Hayward several invitations to visit a shop he kept on Water Street. Several weeks after their first meeting, Hayward visited Goodyear's shop. Goodyear showed Hayward some samples of his work, ornately decorated cloths and aprons and so forth.

Hayward turned them over in his hands. They were smooth and highly decorated. Some were cream colored, others almost transparent, still others black as coal. Hayward said at last, "They're better than anything anybody else is making."

A broad smile broke across Goodyear's face. He insisted that Hayward take some of the samples home. Hayward did so, and when he got back to his factory, the unsentimental ex–stable operator submitted Goodyear's finery to the same rugged test as he gave his own products—he took them outside and nailed them to a sun-exposed outer wall. Within a short time, Goodyear's elegant handiwork slouched on their nails like guilty schoolboys, and Hayward knew Goodyear hadn't unlocked the secret either.

Goodyear, eager to continue his experiments, kept up the pressure on Hayward to sell his mill. Hayward began to relent. His sales were going nowhere. By the middle of 1838, what little reputation rubber still had was gone. Hayward needed cash. In September, the same month he first permitted Goodyear to enter his factory, Hayward agreed to sell it to Goodyear in exchange for $800 per year in salary, with Hayward staying on as foreman. Hayward also agreed to disclose to Goodyear any information he had on rubber. As it turned out, this provision would be infinitely more valuable to Goodyear than the physical factory itself. Where Goodyear planned to get the $800 to pay Hayward is unclear. No doubt he assumed, as he always did, that success and financial rewards were just around the corner.

Sometime prior to the sale, Hayward began experimenting with sulfur as an additive to rubber. Sulfur helped dry the surface of the material. Hayward mixed rubber with sulfur during the compounding phase, or sprinkled sulfur over rubber cloth as it was drying. He placed the drying article in sunlight, a process he referred to as the sun-cure, or solarization. Hayward was always vague about where and how he got the idea, and when he first started using sulfur. Once, he claimed to have first mixed sulfur with rubber back at his stable in Boston in 1834. Other times he said it was as late as early 1838. He even claimed that sulfur appeared to him in a dream: "I dreamt that rubber, lampblack and sulfur was all that was necessary to make rubber."

Just before they closed the sale, Goodyear at last was allowed a complete tour of the Woburn mill. He smelled sulfur and asked Hayward about it. Hayward at first demurred, not sure how much he wanted to reveal. On his next visit to Woburn, Goodyear pressed: "Hayward, I want you to tell me whether you use sulfur, or not. If not, I want you to say so." Actually, Hayward's use of sulfur was less of a secret than he made out. There was no hiding the smell. Among all the chemical smells, sulfur's deep, rotten-egg odor may be the most distinctive, and the most nauseating. And with Hayward's solarization method, the stench of sulfur stayed with the goods long after they left his shop. Samuel Sweetser, a director of the Eagle Company who remained Hayward's landlord at the mill, was alarmed at the stench floating off some rubber cloth he had asked Hayward to make. Sweetser was going to send the cloth to Boston to have Timothy Newell, another investor in the mill, make some shoes. But when the cloth arrived, Sweetser recoiled. The company was having enough trouble selling its wares without trying to convince customers to walk around in shoes smelling of rotten eggs. Sweetser ordered Hayward to stop using sulfur immediately. "My reason was, that I was a director of the company, and I thought it would injure the sale of goods, and it was my interest, being a stockholder, to have such goods as would sell," he later recalled.

When Goodyear pressed him, Hayward relinquished the details. Unlike Sweetser, Goodyear was able to look beyond the objectionable smell to see that Hayward was on to something. Hayward's fabrics, while not impervious to heat, seemed with sulfur to be smooth on the surface. Still enamored of his own nitric-acid process, Goodyear, now with Hayward's help, began experimenting with combinations of nitric acid and sulfur when he took over the Eagle building in the fall of 1838. Clarissa and the children remained in Roxbury at the Norfolk House. When in Woburn, Goodyear boarded in a workers' dormitory on Railroad Avenue that had been built in optimistic expectation of housing silk workers.

Among Goodyear's first acts in Woburn was to urge Hayward to patent the use of sulfur in rubber. Goodyear then purchased the rights to the patent, which would be issued in February 1839. Goodyear's enemies would later use this as evidence that Goodyear was not the authentic inventor of vulcanization, since sulfur was a key to the discovery. Hayward, perhaps understandably, would contribute to this idea from time to time as he saw Goodyear's fame rising and his own contribution lapsing into relative obscurity. But Hayward was at his core a decent, honest soul, and in every important way he acknowledged that Goodyear alone discovered and perfected the process that came to be known as vulcanization. Sulfur, while crucial, was only a start.

Goodyear, for his part, never hesitated to acknowledge Hayward's contribution. "It was through him that the writer received the first knowledge of the use of sulfur as a drier of gum-elastic," he stated plainly in his memoirs. It is possible that Goodyear might have stumbled on sulfur eventually, but there's no telling for sure. For his part, Hayward had a seven-hundred-dollar mortgage on the comfortable white house he'd built on Richardson Row Road, around the corner from the factory. He had a wife and two children. He had had enough of ownership and all of its vast uncertainties.

In the little mill, Goodyear and Hayward began optimistically experimenting with mixtures of rubber, sulfur, and other ingredients, in varying quantities. Goodyear started using white lead, first to improve appearance and then because lead seemed to accelerate reactions with sulfur. Sheets of rubber that emerged from his mill during this time, once dried in the sun, seemed to him to have been "cured"—they were smooth and hard, and seemingly impervious to heat and cold. Goodyear, more excited than ever, began spouting off once again that the great mystery was solved. His overconfidence led to the last and greatest fiasco preceding his actual breakthrough. Goodyear traveled to Boston with samples of his fabric and succeeded in wangling a contract with the United States Post Office for 150 all-weather mailbags. Much rode on the success of this contract. It's a testament to Goodyear's salesmanship that he managed to interest anyone in rubber goods, after all the celebrated failures. And what could better publicize the wonderful properties of his new compound than bags for the men who completed their rounds in rain and sleet and snow? If he could keep the U.S. mail dry, Goodyear knew, the positive publicity might be enough to resurrect both himself and rubber. Goodyear and Hayward spent weeks producing mailbags made of rubber treated with both the nitric-acid process and sun-cured sulfur. Goodyear added white lead and other coloring agents, and soon produced a batch of exquisite mailbags, which he embossed with fancy decorations. So confident was he of success that he produced the bags not from rubber-coated fabric, but from sheets of solid rubber. He professed to everyone who would listen that these products would ensure his financial success.

When the mailbags were finished, Goodyear set them aside and visited his family in Roxbury for a few weeks. When he returned to Woburn, he discovered to his dismay that the bags, like every other brilliant promise before them, had failed. The bags were still smooth and dry on the surface, but the insides had turned to goo. Goodyear had carefully hung the bags by their handsome rubber straps over nails on a wall in the mill. The straps still dangled from the nails, but the bags had sunk to the floor under their own weight. Goodyear faulted himself for being too concerned with appearance; he wondered whether the coloring agents had harmed the products. For Goodyear, this failure conjured memories of the shoe debacle back in New Haven. Only this time was worse. And it had been an expensive mistake as well. Instead of ensuring his financial comfort, the mailbags put him back in the too-familiar territory of poverty. His money from the sale of the shoe patent and the nitric-acid licenses was now all but gone. His family's brief respite from want was at an end.

"The misfortune and disappointment of the writer by this occurrence was indeed serious," he wrote. "He was not only reduced to extreme poverty, with a large family to provide for; but, if he continued his experiments, he could no longer expect the countenance or sympathy of his friends or acquaintances, as he had already spent four years in fruitless attempts to make improvements in the manufacture that had proved so ruinous to the community, having, as was generally known, applied himself industriously to his experiments during the whole time, doing nothing else. It was generally agreed that the man who could proceed further in a course of this sort, was fairly deserving of all the distress brought upon himself, beside being justly debarred the sympathy of others."

In the wake of the disaster, Goodyear was forced to give up the Eagle mill. His stretch in Woburn seemed to have come to a fruitless end. Once again, his friends and acquaintances urged him to put this foolishness aside, for the sake of his family. Once again, he responded to failure in the only way he knew, by pushing ahead. He restlessly turned over in his mind what could have gone wrong. Why had only the surface been hardened? While his family remained in Roxbury, Goodyear had to visit Woburn a number of times to attend to the details of closing up his business. While he was supposed to be engaged with business matters, he found himself re-creating experiments, looking for an answer. Goodyear was not finished with Woburn after all. All at once his signal breakthrough came. It involved rubber, sulfur, and a hot stove. Quite by accident, Charles Goodyear was about to discover vulcanization.

6

BREAKTHROUGH

Few scientific discoveries have occasioned greater speculation, embellishment, and flat-out fabrication than has Charles Goodyear's great discovery in early 1839.

The stretch of East Woburn where Goodyear once lived looks today like a thousand other busy, unremarkable suburban strips: a couple of branch banks, a Friendly's ice cream restaurant, a convenience store, and the incessant rumble of Interstate 93 nearby carrying commuters into Boston. If you look closely enough, though, the present flakes away like layers of paint and you begin to see the outlines of the village of East Woburn as it was in Goodyear's day. The clues are obscured now, but not all of them are gone. Railroad Avenue still marks its east-west course as a main thoroughfare, though the road has long since been renamed Montvale Avenue. The railroad tracks themselves still slice a north-south course past Walker's Pond. Overlooking the pond, at the corner of what was Railroad Avenue and Central Street, stood Elizabeth Emerson's boardinghouse. The house is gone; in its place is an elementary school named for Charles Goodyear. On the side of the building, with unintended irony, students have painted a large mural of the blimp flown by the Goodyear Tire & Rubber Company—the company Charles Goodyear never knew but for which he is now perhaps best known. Just across Central Street a bank occupies the lot of a house where Goodyear lived after moving to Woburn full-time. But a few doors east you will find intact the four-unit row house, built originally to house silk workers, where Goodyear frequently stayed during his early visits to the town. Around the corner, on Washington Street, stands Nathaniel Hayward's sturdy little white house, built in 1837, with a broad front porch supported by six posts. Goodyear spent a lot of time in this house. On the south side of Washington Street the parking lot of another bank marks the spot where the Eagle India Rubber Company factory once stood. Poke around in the overgrown areas between parking lots and buildings and you can still see the faint crease of the canal, now grown in.

The entire area is but a few blocks long, easily covered in a brisk, twenty-minute walk. Somewhere in this old village is the answer to an enduring mystery—the exact location of Charles Goodyear's discovery of vulcanization. The known facts are these: Sometime during the winter, probably in February or March, Goodyear accidentally dropped, placed, or spilled a quantity of rubber mixed with sulfur and white lead on a hot stove. When he retrieved the sample he discovered something remarkable: the sample had not melted, as he expected, but instead had hardened to the consistency of leather. Here, at last, was the final clue, the missing piece to the puzzle: heat. Goodyear and others routinely used heat to soften rubber for mixing, but nobody had ever thought to use heat as a means to protect rubber products once they had been made. Heat, after all, was the great enemy of raw rubber. Using heat to protect against heat seemed as illogical as dunking an object in water to prevent moisture damage. And yet, as this fortuitous accident had shown, something in the combination of rubber and sulfur turned heat from an enemy into a friend. Goodyear could scarcely believe his eyes.

Beyond the few details known about the discovery, all is conjecture, some of it educated, much of it not. Fault for this must lie primarily with the inventor himself. Goodyear's privately published two-volume memoir and paean to rubber, *Gum-Elastic and Its Varieties*, is by turns charming, prosaic, scholarly, clumsy, detailed, elusive, engaging, and maddening. It is never so maddening as when Goodyear describes the great discovery of his career. Over the course of 625 pages Goodyear spares no quantity of ink in describing hundreds of uses for vulcanized rubber. He reveals that vulcanized rubber is the perfect material for a baby jumper, gymnastic rope, clarinet, hot water bottle, syringe, pessary, boat, stocking, self-inflating pontoon, mattress, suitcase, fruit packing case, and envelope. He argues passionately and in great detail for rubber wallpaper, maps, globes, backgammon boards, and sportsmen's pantaloons. And yet the account of his discovery, the incident that gave his life its meaning, justified his suffering, and ensured his immortality occupies this one elusive, elliptical paragraph:

> While on one of the visits above alluded to, at the factory at Woburn, and at the dwelling place where he stopped whenever he visited the manufactory at Woburn, the inventor made some experiments to ascertain the effect of heat upon the same compound that had decomposed in the mail-bags and other articles. He was surprised to find that the specimens being carelessly brought into contact with a hot stove, charred like leather. He endeavored to call the attention of his brother, as well as some other individuals who were present, and who were acquainted with the manufacture of gum-elastic, to this effect, as remarkable and unlike any before known, since gum-elastic always melted when exposed to a high degree of heat. The occurrence did not at the time appear to them to be worthy of notice; it was considered as one of the frequent appeals that he was in the habit of making, in behalf of some new experiment.

The details omitted in this 161-word passage are monumental. Where, exactly, was he? Was his brother present at the actual discovery, or merely present in Woburn at the time? What does Goodyear mean by "carelessly brought into contact with a hot stove"? What stove? For how long was the material exposed? What was the date? In keeping with an age less self-revelatory than our own, Goodyear apologizes for bothering us with personal details at all, certain that we will want to hurry on to the truly compelling aspects of the book—such as his clunky, second-hand rendition of the history of rubber. He would have done posterity a great favor by bothering us with more personal details and less history, but there's nothing to do about that now.

In the absence of precise facts, a cottage industry of myths has grown up around the discovery. One persistent story has Goodyear huddled around a potbellied stove in a general store, talking in animated fashion with a group of villagers. According to the story, Goodyear, gesticulating to make a point, slapped a rubber sample he had in his hand onto the hot stove. Looking down, he was shocked to find that the piece, instead of melting, turned leathery and dark. Another story, a slight variation of the first, has Goodyear discovering vulcanization in the midst of a heated argument. The story was first put forth in 1865, five years after Goodyear's death, in an article by James Parton in the *North American Review*: "He was expiating in his usual vehement manner, the company exhibiting the indifference to which he was accustomed. In the crisis of his argument he made a violent gesture, which brought the mass into contact with the stove, which was hot enough to melt India rubber instantly; upon looking at it a moment after, he perceived that his compound had not melted in the least degree! It had charred as leather chars, but no part of the surface had dissolved. There was not a sticky place upon it. To say that he was astonished at this would but faintly express his ecstasy of amazement."

Other tales have Goodyear dropping a sulfur-tipped match on some rubber; wearing a rubberized apron, spilling sulfur on it, and leaning against a stove; spilling sulfur by mistake into a pot of boiling rubber. An account still current on the Internet casts Goodyear as a peddler who happened upon Woburn while traveling from town to town trying to sell rubber goods. When residents laughed at his haggard appearance, he grew angry and hurled a chunk of rubber against a hot stove, where it stuck and instantly hardened. Another version has Goodyear spilling some mixture onto a stove and being so revolted by the smell that he heaved the mess out the window into the winter night. He was shocked to find it still pliable in the morning. As if Charles Goodyear could still be offended by a bad smell. There is no good evidence to support any of these accounts. Beyond that, they face a serious chemical problem; for any of them to be true, Goodyear's wad of rubber would have had to vulcanize almost instantaneously. In fact, the process is much slower. Even a slender strip of rubber left on a stove at, say, 260 degrees, would require an hour, perhaps longer, for the chemical reaction to take place. Clearly, Goodyear dropped a piece on or in a stove and did not discover it until some time later.

The most attractive myth puts Clarissa at the center. After years of ill fortune and financial hardship (the story goes) Clarissa finally rebelled, forbidding her husband to conduct any further rubber experiments. She ordered him to find a reliable job. A chastened Goodyear kept experimenting on the sly. One day, hearing her returning footsteps, he shoved a batch of rubber into the nearest available hiding place—Clarissa's oven. When he retrieved it, *voilà* vulcanization! Unfortunately, Clarissa was still living in Roxbury when Goodyear made his discovery; she and the children would not move to Woburn until several months later. Even so, the story holds considerable appeal, if only because it finally invests Clarissa with the gumption to tell her husband off—something that even a casual observer of the Goodyear saga longs to do for her.

Serious historians have long focused on the reference to the "dwelling place where he stopped whenever he visited the manufactory at Woburn" as the likely sight for the actual discovery. The reference would make little sense unless it was central to the event. Whether anyone else was actually present when the rubber met the stove, or whether he simply ran out to tell people "present" in Woburn at the time, is unclear. Ellen, his eldest child, who accompanied Goodyear during many of his travels, later claimed to have been nearby when the discovery occurred. And of course Goodyear refers to his brother, Nelson. Getting a fix on the dwelling place is difficult, since the peripatetic Goodyear lived in no fewer than five houses during his three-year, on-again, off-again stretch in Woburn.

For much of the twentieth century, historians believed that the discovery took place in an ungainly, odd-shaped, two-story wooden structure called the Badger House on Railroad Avenue, the main thoroughfare through East Woburn. The house languished in neglect through much of the twentieth century and was finally demolished in 1972, over the cries of historians and editorial writers certain that the wrecker's ball destroyed an important piece of history along with the house. The *Woburn Daily Times* thundered: "The destruction of the Goodyear Home in East Woburn was a callous, wanton act by its owners and an indictment of indifference on the part of all Woburnites." *Rubber & Plastics News*, a trade journal, lamented that "the spot in America where one of the world's greatest industries was born is now a vacant lot."

As it turns out, the eulogies were probably misplaced. The Goodyears did indeed live in the Badger House for more than a year, and Goodyear undoubtedly conducted many experiments in the kitchen. But Goodyear spent his early months in Woburn essentially commuting from Roxbury, staying in Woburn a few days at a time. He didn't move his family full-time into the Badger House until May 1839, at least two months after his vulcanization discovery. Tom Smith, a current Woburn town official and local historian, dug up old real estate records, and compared the records with testimony from Woburn residents taken during the court battles that would occupy much of Goodyear's later years. In his self-published 1986 book, *Goodyear: The India Rubber Man in Woburn*, Smith argues persuasively that Goodyear most likely boarded during those months in the row house on Railroad Avenue built to house the expected influx of silk farm workers. With Goodyear needing a place to stay and the brand-new dormitory in want of tenants, it would have been a logical fit. The house, just a quarter mile from the factory, was run by Timothy Newell, one of the Eagle Company directors. Newell is known to have helped Goodyear on some of his experiments. Short of conclusive proof about the location of the discovery, Smith's theory stands as the soundest to date. And it holds out the tantalizing possibility that the site of Goodyear's discovery may not have been lost to the wrecker's ball after all. The row house was built in 1837, a year before Goodyear came to Woburn. Two centrally located chimneys provided fireplaces for all four units. Today, the white, wood building is privately owned and occupied. It's the sort of place you drive by without looking at twice, an unremarkable structure. It is simply another building on Montvale Avenue. There are no markers, nothing to indicate that anything special might have happened here, nothing to ensure the house against destruction. And yet fate has given the town of Woburn, which thought once that it had lost forever the site of Goodyear's great discovery, a second chance to preserve history.

When Goodyear stumbled upon the secret of "fire-proof gum," as he called it (the term "vulcanization" would be developed by a competitor several years later), nobody wanted to hear about it. Even Goodyear's closest friends and associates had had enough of his solutions and pronouncements. For all the travails he had endured prior to 1839, the years immediately after would prove to be the most frustrating and heartbreaking of his life. At last, he had the answer, but nobody was asking the question. Friends and associates had seen him through the magnesia failure, the nitric-acid mishap, the sulfur mailbag fiasco. What could this new "discovery" be but another dead end? Charles Goodyear knew this was different. He sensed from the moment he yanked the spilled sample from the stove that he had stumbled onto the one true secret.

Much has been made over the years of the fact that Goodyear's discovery was accidental, the inevitable conclusion being that he was simply a lucky rube who stumbled upon greatness. Contemporary humorist Bill Bryson captured this sentiment with this amusing assessment: "Goodyear personified most of the qualities of the classic American inventor—total belief in the product, years of sacrifice, blind devotion to an idea—but with one engaging difference. He didn't have the faintest idea what he was doing." This stereotype has been current for generations, if not always expressed with Bryson's characteristic wit.

The fundamental flaw in calling Goodyear a fortunate rube is that human discovery is largely the story of accident converted into opportunity. Alexander Fleming, the Scottish bacteriologist, wasn't looking for penicillin when he noticed a strange mold growing in an old culture dish. But where a hundred other men would have reached for a scrub brush and trash can, Fleming's peculiar genius was to recognize potential in this humble growth. The result was the birth of antibiotics, millions of lives saved, and a Nobel Prize for Fleming.

Goodyear himself was sensitive to the suggestion that he had merely chanced into greatness. He would write in his customary third-person voice: "It may be added that he was many years seeking to accomplish this object, and that he allowed nothing to escape his notice that related to the subject. Like the falling of an apple, it was previously suggestive of an important fact to one whose mind was previously prepared to draw an inference from any occurrence which might favor the object of his research. While the inventor admits that these discoveries were not the result of *scientific* chemical investigations, he is not willing to admit that they were the result of what is commonly termed accident; he claims them to be the result of the closest application and observation."

Scientists have known since Goodyear's time that rubber is composed of carbon and hydrogen, in the proportion of five carbon atoms to eight hydrogen (C_5H_8). There is nothing extraordinary about this. At first glance, rubber's molecular physique looks positively puny next to other compounds. Even humble table sugar ($C_{12}H_{22}O_{11}$) looks imposing enough to kick sand in rubber's face. This does little to explain where rubber gets its bounce, its kick, its snap. So what makes rubber so special?

The answer would not come until the 1920s, when a German chemist named Hermann Staudinger discovered that a rubber molecule was not merely a single unit of C_5H_8. It was instead an *enormous chain* of C_5H_8's strung together in single file. No fewer than twenty thousand of them comprised a single molecule. Compared with ordinary molecules, these *macromolecules* are enormous, slithering serpents.

Suddenly it all becomes clear, why rubber pulled or dropped from a table onto the floor behaves so differently from, say, a sugar cube. A piece of rubber the size of your finger is essentially a nest of millions of long chains, hopelessly twisted and kinky, so tightly wound that not even water can get through. Hold the piece by its ends and stretch it. In a flash you have straightened out millions of long, crooked chains. Let go and the rubber snaps back. Squeeze the rubber between your fingers and the same process happens, only in reverse.

Rubber molecules, while deeply intertwined, are not connected with one another on a molecular level. This is essential to rubber's pliability, but it is also the source of the problem that confounded Goodyear and others for so long. When rubber is heated, the excited molecules relax their grip on one another and slither apart into sticky goo or liquid. In extreme cold, the molecules freeze and crystallize like water into ice.

But how, exactly, did sulfur and heat solve the conundrum? Was it a chemical or physical reaction? Even with their new understanding of macromolecules, scientists weren't sure. In 1936, the *Vanderbilt Rubber Handbook*, one of the industry's most venerable publications, would state, "Theoretical discussions of vulcanization are of little value to the compounder. Practically everything that is known today about compounding has been found out by empirical methods."

Only in the last few decades have scientists come to understand just what happened that day in Woburn. In the rubber macromolecule, a certain percentage of the carbon atoms are unsaturated—meaning they have unfulfilled potential to join with other atoms. In the absence of a better offer they simply form double bonds with other carbon atoms in the same molecule. When sulfur is added at room temperature, the sulfur atoms and carbon atoms sit on their hands like college freshmen at their first mixer. But look what happens when you turn up the heat and, in Goodyear's case, add some white lead as an accelerant. It's like adding salsa music and a keg of beer to the mixer. The wallflower carbons and sulfurs transform into extroverts, on the prowl for partners. Since the sulfur atoms can link with carbons in more than one rubber molecule simultaneously, suddenly the rubber macromolecules connect like ladder posts joined by hundreds of sulfur rungs.

The process is called cross-linking. The more sulfur you add, the more cross-links you'll produce. One sulfur connection per two hundred rubber monomers (the individual hydrogen-carbon units) is sufficient to vulcanize rubber. More sulfur will make the resulting product more durable, but less flexible. Hard rubber, or vulcanite, developed a decade after Goodyear's discovery, is simply rubber to which a great deal of sulfur has been added.

Of course, Charles Goodyear understood none of this, only that he had stumbled upon something magnificent. In the days immediately following his discovery, Goodyear furiously mixed batches of rubber and sulfur and set them in, on, or near any available oven. But try as he might, he could not repeat his success. He assumed in the glory of his Eureka! moment that he would soon be rolling finished products off an assembly line. But vulcanization is a delicate process. Modern manufacturers vulcanize natural and synthetic rubber by a variety of methods—in closely regulated steam pressure ovens, by heated gas, in hot air tunnels, by microwave, in molten salt baths, or by radiation beam. Even under the most sophisticated conditions, mistakes occur. Improper heating can result in blistering, uneven vulcanization, or in complete failure. It is nothing short of astounding that Goodyear, hovering over a stove emitting uncertain temperatures on an irregular mass of rubber, impregnated with an imprecise amount of sulfur, for an indistinct period of time, managed to vulcanize rubber at all. From here, endless experiments would be required to learn how to repeat the process under controlled conditions. Without reliable repetition, the process was useless.

The next two or three years would be filled with intense frustration for Goodyear—a mishmash of exultation and the trembling joy of feeling he was close to the mountaintop, combined with a depressing sense of how far he had yet to go. During this period Goodyear's greatness fully emerged. After vulcanization became universal, others would claim that they had developed the procedure first, that they had mixed sulfur and rubber in some laboratory years before Goodyear. But none of these claimants could demonstrate that they had found the crucial combination of rubber, sulfur, and heat. None of them invested the agonizing weeks, months, and years to perfect the process, to bring it under control, to make it useful. Putting a harness on rubber exacted a monumental cost in time, dedication, and sacrifice. Only one man had the obsession to persevere. Charles Goodyear.

By early 1839, money Goodyear had earned from licenses was gone. About this time he made a gesture that was either quite foolish or quite noble (perhaps both). When a French manufacturer offered to buy the secrets to his nitric-acid process, Goodyear declined to sell, saying that he was on the verge of perfecting a process that would render the acid-gas solution obsolete. He did not wish to sell an outdated process when he would surely soon have his fireproof process complete. Much of his licensing money had gone to pay off old debts. Goodyear was irresponsible in acquiring debt and tardy in paying it off—but he was not a cheat. Whenever he could, he paid old debts. A chunk of his cash had gone into paying Hayward and setting himself up at the Eagle factory in Woburn.

In May, Clarissa and the children moved out of their comfortable home at the Norfolk House in Roxbury and joined Goodyear in Woburn. The family moved into the Badger House, at the corner of Railroad Avenue and Central Street. It was a full house. In addition to the immediate family, Goodyear's aging parents, Amasa and Cynthia, and his brothers Nelson and Henry were frequent boarders.

As the Goodyears set up house, Charles tried desperately to reproduce the magic. The breakthrough seemed so close that he could almost taste it—five years of frustration, privation, and false starts were, at last, about to come to an end. But he could not produce a uniform piece of rubber. Some samples were not vulcanized at all; others were hard in places, soft and tacky in others. Some pieces blistered or emerged burned to a crisp. He could not repeat his results. Weeks passed, then months, with no results. Before long, he had no fuel for his fires. He began wandering the streets of East Woburn, asking smiths and millers if he might use their fires when their work was finished for the day. John McCaghil, a twenty-six-year-old Irish immigrant to East Woburn by way of New Brunswick, worked as a smith at the Alexander Richardson & Company Saw Mill, not far from the Eagle building. McCaghil tempered saw blades in a large furnace built of fire brick, ten feet long and forty-six inches wide. The furnace was too attractive for Goodyear to pass up. He would appear at Richardson & Company toward the end of the work day, clutching rubber samples and inquiring politely if he might use the fire when McCaghil was finished. McCaghil could think of no logical reason to say no, since the furnace needed time to cool down anyway.

Goodyear shoved batches of two or three samples at a time into McCaghil's oven on metal sheets. If he had some small success with little pieces of rubber cloth, larger items were hopeless—they required extra heat to vulcanize, but the extra heat would invariably blister the surface of the rubber. Once in a while, though rarely, a section of rubber would emerge tantalizingly, agonizingly smooth, but temporary exhilaration always yielded to disappointment and frustration.

Charles Whittemore, another East Woburnite, ran a dye factory. Goodyear persuaded Whittemore to allow him to run pieces of rubber through the cylinders used for finishing cloth. Whittemore reluctantly agreed. When Goodyear ran his pieces through, they left behind splotches of melted rubber on the rollers, which in turn left spots on Whittemore's fabrics. Goodyear also talked James McCracken, Whittemore's fireman, into letting him place pieces of rubber on top of the boiler or hang them next to the exhaust pipe of the steam engine. At other times he would open the furnace door, holding items over the flames on the blade of a shovel, like some kind of pathetic overgrown Boy Scout toasting marshmallows. McCracken grumbled that each time the door opened, heat escaped, making it that much harder to maintain a steady fire. Whittemore hated the splotches on his fabrics. Goodyear apologized but kept right on pressing.

Amid this compact community of farmers, laborers, shopkeepers, and boardinghouse operators, Charles Goodyear was seen as an oddity, a curiosity, a crank. As Samuel Sweetser, an erstwhile Eagle partner, noted, "It was about death to a man's credit to be engaged in India rubber business." For more than two years this odd man and his family attached themselves to East Woburn, offering to the community only their burdens of poverty and need. The forbearance the Woburnites showed to this strange, persistent man who was not, after all, a native of the area is one of the more remarkable chapters in the Goodyear saga.

7

IN THE DARKEST HOURS

About the only uncomplicated thing in Charles Goodyear's life was the clarity of his mission. He never wavered, never thought seriously of giving up. The worse things became, the stronger his resolve grew. He did not seek poverty, yet neither did he shy away from it. He explained in his memoirs that God had chosen him to solve the rubber mystery and that he, Goodyear, was merely "the instrument in the hands of his Maker." His self-certainty on this score allowed him to excuse many of his own failings—his carelessness with other people's money, for example, and his inability to provide for his family. But it also gave him a rare and awesome strength.

While Goodyear struggled to bring his discovery under control and to hold the torn rags of his life together, Thomas Hancock continued building a small empire on imperfect rubber. His prudent, cautious avoidance of rubber's flaws seemed to be paying off. Hancock had his share of setbacks, to be sure, but even these helped him solidify his control of Charles Macintosh & Company and, hence, the British rubber industry. In the middle of a summer night in 1838, fire swept through the company's Manchester factory. Large quantities of raw rubber and volatile solvents produced a noxious, ferocious blaze. Nobody was working at the time the fire started, but employees rushed back to try to put it out and to save equipment. Several were crushed when large pieces of machinery crashed through fire-weakened floors. Hancock was in Scotland visiting Macintosh (by now in semiretirement) when the news reached him. He hurried to Manchester to find the brick walls, stone stairs, and chimney intact, but much of the machinery and virtually all of the stock was destroyed. Hancock took charge of the disaster scene, organizing the cleanup effort and almost immediately setting plans to resume manufacturing. Two large steam engines had escaped serious damage, and after repairs to the machines and the building, the factory reopened within a matter of weeks, thanks in no small part to Hancock's abilities as a field marshal.

By 1840 business had recovered so robustly that the company expanded its Manchester factory. Workers turned out more than three thousand square yards of double-texture cloth per day. The company operated two stores in London, at 46 Cheapside and 58 Charing Cross. The stores stocked raincoats, air cushions, inflatable pillows, and other goods, which sold steadily enough to attract imitators. An advertisement in a London paper that year warned consumers against "spurious articles" and urged them to check closely to ensure they were buying the real thing. "None are genuine which have not the autograph C. Macintosh & Co. engraved with the Royal Arms on the lining."

In stark contrast to Hancock's position as captain of his industry, Charles Goodyear's industrial world had shrunk to the confines of Clarissa's kitchen. His endless pleas and messy experiments had strained the patience of even the remarkably patient artisans of Woburn, who increasingly turned him away when he showed up asking to use their fires. Goodyear had no choice but to restrict his experiments to the territory of the one person whose patience reigned supreme, his wife. When Clarissa's cooking was done for the day and the rest of the family was in bed, Goodyear converted the kitchen into a makeshift laboratory. He baked batches of rubber in Clarissa's bread pans. He dangled patches in front of the spout of the tea kettle as it boiled. He suspended pieces from the handle of the teapot over the opening of the kettle. He toasted strips in dying embers in the stove.

More than ever, Goodyear needed a steady and controllable fire, and a chamber large enough to heat pieces larger than small strips. Somehow, he convinced Samuel Sweetser, landlord of the defunct Eagle plant, to allow him to build such an oven at the factory. Then he convinced several residents to help build it. Goodyear, his father, and his brothers Nelson and Amasa Jr. dug the foundation. Daniel Burbank, a mason, laid the bricks. Alden Moore did the carpentry, and Isaiah Wadleigh the ironwork. Wadleigh had allowed Goodyear to use his own furnace from time to time and was probably eager to help, if only to keep Goodyear and his rubber samples away from his own door. The completed oven was about ten feet long and seven feet wide. The oven compartment was lathed and plastered on the inside, with a cast-iron stove on top. The outside of the oven was brick, and it had a door covered with zinc. Wood was added on the outside of the building, but the door through which objects were placed in the oven was on the inside. Whatever they had been promised in the way of payment for their labor, the workers received little for their efforts other than samples of various rubber goods Goodyear had been making, as well as promises of future payment.

With his oven completed, Goodyear proudly invited Luke Baldwin, the old foreman from the Roxbury India Rubber Company, out for a test run. James McCracken, the tolerant fireman from Whittemore's shop, was on hand when Baldwin visited. Goodyear placed a long roll of rubber cloth in the oven and shut the door. After a time, which Goodyear no doubt filled with excited details of his progress, what emerged from the oven was such a sad, burned-up, pathetic-looking mess that McCracken and Baldwin burst out in spontaneous laughter.

Baldwin, shaking his head and wiping away tears of laughter, nodded at the oven and said, "It's good for nothing." Goodyear, normally the implacable optimist in the face of failures and humiliations, grew visibly upset. To Goodyear it seemed his old friend was laughing not at the single rubber sample, but at all of his efforts and struggles. Even his erstwhile compatriots, the last remaining core of rubber believers, seemed to believe that Goodyear's struggles had been reduced to a joke. The only record is McCracken's dry, understated reflection that Goodyear "did not seem very well pleased with what Mr. Baldwin said at the time."

Not everybody laughed off Goodyear's breakthrough. Among the few important supporters amid the sea of doubters was Benjamin Silliman Jr., a prominent Yale chemist and geologist who had befriended Goodyear during the mid-1830s when the latter returned to Connecticut from Philadelphia. Silliman, widely respected for his developments in uses for petroleum, had provided testimonials authenticating some of the inventor's early rubber experiments. In October of 1839, Goodyear visited Silliman in New Haven, hoping for another testimonial. After examining some samples, Silliman wrote of Goodyear's rubber: "I can state that it does not melt, but rather chars, by heat, and that it does not stiffen by cold, but retains its flexibility in the cold, even when laid between cakes of ice." Coming from an impeccable source, the testimonial, dated October 14, 1839, would later provide a crucial piece of evidence bolstering Goodyear's claims as the developer of vulcanization.

For now, though, Silliman's words could do little more than lift Goodyear's spirits. Goodyear still was having trouble vulcanizing strips of rubber with any consistency. Even more vexing was producing usable objects made of the substance. After vulcanization, rubber loses its signature ability to be molded into various shapes. It can't simply be melted down. That's why, even in our own time, worn-out tires remain an environmental nuisance. Despite recent breakthroughs in "de-vulcanizing" rubber with solvents, vulcanization is essentially a one-way street. For

Goodyear (and every rubber manufacturer after him) this phenomenon meant that objects made of vulcanized rubber had to be molded and assembled prior to vulcanization. It was difficult enough to vulcanize a piece of rubber-coated fabric. Trying to vulcanize, say, a pair of shoes evenly proved to be devilishly hard.

The process was also expensive, time-consuming, and wasteful. It was one thing to ruin some stray scraps, quite another to send a batch of newly finished shoes or hats to their doom. One day Goodyear walked into the shop of shoemaker William Trotman with assorted pieces of rubber fabric he wanted sewn into several pairs of shoes, as well as a rubber canteen he wanted stitched. Trotman examined the pieces to be made into shoes and laughed the laugh Goodyear had heard so many times before.

The inventor said, "Would you be astonished if I can produce India rubber that will take the place of kid and leather?"

"Yes. Astonished indeed."

"I will," Goodyear said. "One day I will."

Trotman remained skeptical, but made the shoes for Goodyear as ordered. When they were finished, Goodyear picked them up and left with a promise to pay soon. A couple of days later, after another round of failed experiments, Goodyear returned, not with payment, but with an order for two more pairs. Trotman made them. Again, Goodyear promised to pay in a few days. He didn't. But as irresponsible as he was with money, Goodyear was a man of honor after his own fashion. Three months later, Goodyear returned to Trotman's shop bearing a tablecloth, which he handed over to the shoemaker.

"Keep this as security until I pay you for the shoes," Goodyear said. Twelve years later, Trotman still had the tablecloth.

Herbert Beers, another Woburn shoemaker, spent four weeks making a much larger batch of boots and shoes for Goodyear. Beers proudly delivered the shoes to Goodyear and Hayward at the factory, only to watch nervously as Goodyear ordered every pair into the oven. Goodyear and Hayward stoked the furnace with wood. After half an hour a smoky stench seeped from the oven. They opened the door, and smoke billowed out, making them gag and fall back. The wooden lasts inside the shoes, used to keep them in shape during vulcanization, had ignited under the intense heat. The burning wood in turn ignited the rubber. Goodyear, Hayward, and Beers doused the mass and stepped back to catch their breaths.

Goodyear was the first to speak. "Well, there's no help for it," he said matter-of-factly. "Herbert, make us some more shoes."

Beers looked down at the revolting mess that represented a month's labor. "But we've just lost the whole lot," he stammered. "Hadn't I better make fewer pairs this time?"

Goodyear waved him off. "Don't worry about the cost. If I accomplish my object, it will be a great benefit to me financially."

Beers spent three weeks putting together more shoes and boots from rubber fabric supplied by Goodyear. This time, Goodyear and Hayward tried heating the oven more slowly, keeping the heat just low enough to avoid burning the lasts. The result was a significant improvement over the previous attempt's total loss, but the shoes had not been fully vulcanized. Beers produced yet another batch of shoes. This time, Goodyear filled them with sand to replace the wooden lasts. The sand wouldn't ignite, but it did become hot enough to blister the shoes from the inside.

Some of Goodyear's experiments did little but reinforce the general view that he was a lunatic. Long a believer in the military possibilities for rubber, he rigged up some prototype naval mines. Goodyear, Hayward, and Beers (drawn now into Goodyear's sphere of helpers) took a large jug filled with gunpowder down to a mill pond attached to the canal. They attached a long hose made of rubber cloth to the mouth of the jug. Then they fed a fuse down the center of the hose into the jug, and submerged the jug in several feet of water. With the glee of schoolboys on a prank, they lit the fuse and waited two minutes while the flame crawled down the length of the slow-burning fuse. Suddenly the pond erupted; a geyser of water shot up from the center of the pond, spraying water for ten yards in every direction. A delighted Goodyear pronounced the experiment an unqualified success. Later, Goodyear and Hayward assembled a small inflatable boat from rubber cloth. With Beers's help, they dragged it out to the canal, loaded themselves aboard, and flipped themselves over into the water.

By the winter of 1839–1840, Goodyear still had not found a way to reliably reproduce vulcanization. He had fallen drastically behind in his payments to Sweetser for rent of the factory. Sweetser later recalled that Goodyear's credit had fallen so low that he "could not get trusted for an ounce of tea." In the small village of East Woburn, Goodyear inevitably bumped into Sweetser, often in a country store not far from the factory. He continued to extol the virtues of vulcanization. He staged demonstrations for his landlord, hoping to entice him as an investor. He would take along strips of rubber cloth he had successfully vulcanized. To emphasize their durability, he wrapped the strips around his finger, then held his finger to a lighted candle. The rubber withstood the heat without melting. Sweetser admired Goodyear's handiwork, but he viewed those rubber samples with the sort of guarded, arm's-length admiration one feels for a convicted criminal who finds religion behind bars. It was too little, too late. Sweetser had already lost five thousand dollars on rubber, and he wasn't about to become a rubber investor all over again after finally freeing himself from the Eagle fiasco. Besides, a small strip of rubber cloth was just a small strip of rubber cloth. Sweetser doubted whether Goodyear could reproduce the effect on larger materials. Sweetser would have readily traded Goodyear's parlor tricks for a month's rent on the factory.

Nearly a year had passed since his discovery, and Goodyear was little closer to perfecting vulcanization. Just as Amasa and Cynthia had done in Naugatuck, the Goodyears kept a vegetable garden for food. But as the inventor's fortunes sank, neighbors noticed the Goodyear children digging half-grown potatoes from the field. Though Goodyear's credit was tapped out at every store in town, sometimes a barrel of flour or some other gift would arrive unexpectedly at the door, saving them from starvation.

Elizabeth Emerson, who lived across the street from the Goodyears and had befriended Clarissa and the children, pitied their "extreme destitution." "Their family was sick, and I was called to be with them," she later remembered. "I found they had not fuel to burn, nor food to eat, and did not know where to get a morsel of food, from one day to another, except it was sent to them." She resented Goodyear on behalf of his family, for "experimenting so much on the rubber, and spending all the money he could get on that business, until he was entirely destitute."

Goodyear's frustration manifested itself in a new and terrible fear—that he would die before he could convince the world of what he'd found. Against this ultimate deadline, Goodyear saw no shame in taking charity where he could find it, nor in pawning off the last family possessions. His children's schoolbooks brought five dollars to buy more rubber supplies. "Small as the amount was, it enabled him to proceed," he explained in his memoirs. "At this step he did not hesitate. The occasion, and the certainty of success, warranted the measure, which, in other circumstances, would have been sacrilege."

In his own way, though, Goodyear was a loving father. Ellen, the eldest daughter, always remembered him fondly and recalled that the father regularly read passages of the Bible to the gathered children. And through the darkest days, Goodyear's children seem to have been a particular comfort to him. He was especially fond of William Henry, a happy, sweet-natured toddler now about two years old. William Henry had been given the same first name as his brother, who had died in New

Haven a couple of years earlier. William Henry delighted his father, always managing to draw a laugh, even as heaviness closed in on the inventor from every side. One day, Goodyear returned home and removed his shoes. Before long, William Henry was parading around the room in one of his father's shoes, with his foot inserted not in the conventional place, but through a ragged hole in the toe. The moment, full of pathos and innocence, offered an unintentional commentary on the family's economic condition. For that short time, "Willie" threw the whole dark situation of their lives into relief. Goodyear doubled over in laughter, and soon the rest of the family joined in—things weren't so terrible, it seemed, if they could still laugh.

The winter of 1840 was particularly cold. Snow fell early and often, piling up against the porch, blowing in gusts down Railroad Avenue, forming thick layers over the ice-crusted river, canal, and ponds. Goodyear could be seen tromping through the snow, searching for scraps of wood to heat his home and fuel his brick oven. During one of the frequent storms of that winter, he hit bottom. The family was stuck in the cold house without fuel or food. He approached Sweetser for a loan. But the landlord turned him away. Goodyear decided to try another former Eagle Company director, Oliver Coolidge. He seemed to recall that Coolidge had uttered a kind word to him in passing. Coolidge lived several miles away in North Woburn, up over Wood's Hill and beyond.

With no firmer prospects than the vague remembrance of a kind word, Goodyear wrapped his coat tightly around him and set out in the middle of a snowstorm. He breathed heavily climbing the long hill, stopping periodically to catch his breath by leaning against a snowbank. Wind-borne snow whipped at his face as he trudged along. At last he arrived at Coolidge's door. It took Coolidge a few moments to recognize the gaunt, bedraggled snowman at his door as Charles Goodyear. He welcomed the inventor into his home. As he warmed himself by

Coolidge's cheerful fire, Goodyear produced some scraps of vulcanized rubber from his pockets and began extolling their virtues to the patient host. There was something in the sparkle of Goodyear's eyes that enlivened his pitiful face. Coolidge, charmed, encouraged him to continue. At the end of Goodyear's visit, Coolidge agreed to advance Goodyear six hundred dollars, to see him through the rest of the winter and allow him to feed his family and continue his experiments. Goodyear, overwhelmed by the act, told his benefactor impulsively that the money entitled him to the earnings on any future rubber products. The gesture both touched and amused Coolidge: what possible wealth could ever be generated by the dreamer sitting before his fire in a snowstorm? When the true value of Goodyear's discovery at last became known, Coolidge might well have demanded to collect on Goodyear's rash promise. Certainly, he had more of a claim than did most of the vultures who picked at Goodyear's bones a few years later. But Coolidge recognized the promise as the half-delirious ramblings of a man who for the moment had been delivered from despair.

Goodyear, true to form, spent too much of Coolidge's largesse on rubber and equipment and not enough on his family. He had had some success with vulcanizing larger pieces, though certainly not enough to justify what he did next. He decided to spend the balance of the winter producing "military equipments"—probably including the mines he was so enamored of. The idea was to make large batches of objects to be used as potential specimens, then vulcanize all of them in the spring. All he would get from this venture was another hard lesson about the fickle properties of rubber. When left more than a few weeks, the sulfur-rubber mixture breaks down and makes vulcanization impossible. By spring, Goodyear found himself once again short on cash, with a mass of useless objects to show for a long, bitter winter's labor.

In April 1840, Goodyear and Hayward traveled to Boston together on business. When they arrived, an officer arrested Goodyear on some old debt. Goodyear had no intention of letting the time go to waste. He asked Hayward to send him a rubber bed, two dollars, and a bunch of small pieces of rubber they'd been working on. He wrote to a pair of business acquaintances from his cell, masking his pain and humiliation with strained good spirits, comparing the cell with a hotel room:

Debtor's prison, Boston, Mass.
April 21, 1840

Gentlemen:

I have the pleasure to invite you to call and see me at my lodging on matters of business, and to communicate with my family, and possibly to establish an India rubber factory for myself on the spot. Do not fail to call on the receipt of this, as I feel some anxiety on the account of my family. My father will probably arrange my affairs in relation to this hotel, which, after all, is perhaps as good a resting-place as any this side of the grave.

—Charles Goodyear.

Another letter, written the same day to his old friend William Ely, belies a darker image, showing the grinding effect of poverty and repeated imprisonment on his spirits: "I have fallen into the hands of the harpies, and I do not suppose that they will now let me alone. The amount is small. You recollect the note given me for acid, $50 or $75, on which you paid a part and which was passed into the hands of some stranger. After the current sets that way, a man need not be disappointed or surprised to find himself imprisoned by his own wife."

In early May, Goodyear would again find himself in jail, this time in Cambridge. "Since I wrote you last, I have been imprisoned in Cambridge, on an old execution of $40. Close confinement. No fun in this," he wrote to Ely.

Having nearly exhausted the patience and generosity of Woburnites, Goodyear began casting a wider net. He remembered a friend in Boston. He hoped to garner fifty dollars to travel to New York, where he might convince some prospective investors of the value of his work. Leaving his family behind at the house on Railroad Avenue, Goodyear set off for Boston. He could not afford the train, so he walked the ten miles. When he arrived, he settled into a hotel before setting off the next morning to find his friend. The friend, perhaps stung by past loans to Goodyear or aware of his shaky reputation with finances, rejected him out of hand. Goodyear pounded the streets for a week with no luck. He was an odd-looking nomad, worn, sickly, and haggard. Passersby probably remarked on his ugly coat. It was a drab, light-colored coat made of rubber-coated fabric. The tails had dark, burned splotches on it from Goodyear's experiments.

By Saturday Goodyear was tapped out. He had no money to pay for his hotel room, let alone to travel beyond Boston. Finally, he tried one last house. His hope of reaching New York was gone; now, he simply wanted five dollars to cover his bill. This latest acquaintance slammed the door in his face. Goodyear returned to his hotel, explained his situation, and was unceremoniously evicted (but, fortunately, not jailed). He crossed a bridge over the Charles River, to East Cambridge, where still another acquaintance refused to give him money but at least offered to put him up for the night. The next morning, weary and depressed, with empty pockets and the whole world weighing on his shoulders, he started the long walk back to Woburn.

At last, he gratefully pushed open the front door to his home, happy at least to be with his family. Willie, the toddler, who had been happy and playful when Charles had left for Boston a week earlier, had taken ill. Within a few days, Goodyear's beloved son was dead. Though there is no record of the funeral, it most likely took place at the Second Burial Ground, which served as the town's public cemetery from 1794 to 1845. Goodyear had no money to pay for a coffin, nor even a carriage to take the body on the mile-long uphill journey. It was by now late May, well into springtime. Flowers had unfolded from the earth and buds were bursting on the trees. Neighbors watched as the sad train of mourners left the Goodyear house, wound along Railroad Avenue, past Walker's Pond, over a short bridge spanning the railroad tracks, and up the long, winding road on Wood's Hill. The springtime only seemed to mock the procession.

The circle of those who still believed in Goodyear's quest was growing ever smaller. Now, even his father gave up on rubber. Amasa and Charles had been partners in success and, more often, in failure for almost twenty years. After the collapse of the hardware store, Amasa had eagerly followed his son's new obsession. Amasa and Cynthia lived with them in Roxbury and, for a time, in Woburn. Amasa had accompanied Charles on his first visit to Nathaniel Hayward in Woburn, and had helped build Charles's oven. He was the original dreamer. But now Amasa lost interest in an endeavor that was clearly going nowhere. As poverty engulfed the extended family, he hatched a new plan for getting rich. He had heard about cheap land in Florida. Leaving Cynthia behind with relatives in Connecticut, Amasa headed to Florida with plans to cultivate oranges for export to the north. Amasa Jr., Charles's youngest brother and sometime assistant, required little convincing. His father's talk of sunshine and tropics must have painted a fresh splash of promise against the dingy prospects of staying hitched to Charles in Woburn. Amasa Jr., twenty-seven years old, with a new wife and an infant daughter, had grown weary of the constant failures. He packed up his family and headed for the Florida Keys with his sixty-eight-year-old father.

Charles returned to begging. A friend mailed him seven dollars, along with a stinging rebuke for having neglected his family's needs. Then Goodyear wrote to William DeForest, his brother-in-law in Connecticut. DeForest sent fifty dollars. Goodyear used the money to travel to New York to meet with a New York merchant named William Rider. Rider and his brother, Emory, were successful merchants who for years had been on the periphery of the rubber trade. Goodyear packed up a bunch of rubber samples he hoped might convince them to take a financial interest in his rubber improvements. The trip to New York was important in many ways. For one thing, it was the beginning of the end of his time in Woburn. Although he would continue to live and experiment there off and on for two more years, his attention would increasingly turn to Springfield, Massachusetts, where the Rider brothers would set him up in a rubber factory. The Riders introduced him to Stephen Moulton, a British émigré who had set up a business brokerage at 48 Greenwich Street. The meeting was fateful for both men. Moulton would become a key ally of Goodyear's in England; and he would use knowledge of rubber imparted by Goodyear to become one of England's most successful rubber manufacturers.

Unlike most people he'd spoken with in the past two years, the Riders were impressed with the samples Goodyear showed them. They agreed to finance his experiments and pay living expenses for his family. Goodyear began shuttling back and forth among Woburn, Springfield, and Northampton, Massachusetts, a mill town a few miles north of Springfield, in the western part of the state. The family followed. They lived for a time in Northampton, and returned to Woburn for a short stretch in 1841.

Goodyear was hard at work when heartbreaking news arrived from Key West late in the summer of 1841. Yellow fever, a constant danger in nineteenth-century Florida, had swept the key. This singularly nasty virus is thought to have arrived in the New World from Africa via the slave trade. A few days after being bitten by an infected mosquito, a victim suffers headaches, fever, and muscle pains. Later, sometimes after a deceptive recovery period, the full-blown disease sets in. Liver failure produces jaundice of the eyes and skin (hence the name). Victims suffer severe headaches, internal hemorrhaging, delirium, and death. Amasa Jr. succumbed first, on July 1. Five weeks later the disease took his wife, twenty-four-year-old Melinda, leaving their daughter, Harriet, an orphan. The elder Amasa cradled his infant granddaughter in his arms until she died on August 15. Amasa himself had just four more days to live, surrounded by the specters of his dead family and broken dreams. The citrus plan, like the domestic hardware store, had been at once prescient and tragically before its time. All Amasa reaped was failure, and the cruel order of the deaths only intensified his agony. He died knowing that those he loved had shared the unspeakable price of his dreams.

The news devastated Charles. Not only had he lost his youngest brother and his family, but with the death of his father he lost the closest thing he had to an intellectual compatriot and the primary source of his inventive passion. Men naturally see in their fathers at least a partial blueprint for their own destinies. For Goodyear, the parallels with Amasa were inescapable and ominous. He did not have to look far to see the huge cost his rubber obsession was exacting on his own family, including the death only months earlier of little William Henry. But if Goodyear ever considered abandoning his goal, he did not do so for long. It was simply not in Charles Goodyear's character to give up.

8

THE DASTARDLY HORACE DAY

As he moved his operations from Woburn to Springfield, Goodyear slowly began making breakthroughs in the vulcanization process. Slowly, slowly, he was learning to control the beast. For one thing, his equipment was becoming more sophisticated. He no longer depended on the fickle heat of Clarissa's and Mrs. Emerson's stoves and tea kettles, nor of James McCracken's blacksmith fire, nor even of his own crude furnace at the Woburn factory. With the Riders' money in hand, Goodyear settled into his new factory in Springfield, located along Mill River. Among his first steps was overseeing the construction of a larger, improved oven. This one had cast-iron plates, was six feet square and eight feet tall, and its sides were covered with brick. The fire was set beneath the bottom plate, but pipes diverted heat around the oven. Inside the oven was a rotating horizontal reel. Objects placed on the reel rotated, powered by a waterwheel. The heat diversion and the rotating plate helped to make the heating more uniform.

Still there were setbacks. One day, the reel keeled over, starting a fire that destroyed the reel and an entire batch of goods, and heavily damaged the inside of the oven. With characteristic resourcefulness and resilience, Goodyear took the opportunity to redesign the oven. This time, he had workers install a vertical shaft with iron wheels. Objects rotated end over end, like Cornish game hens in a rotisserie oven. Goodyear also knocked out the iron plate at the bottom of the furnace, which he suspected of radiating too much heat to the lower portions of the oven. Now, with heat relayed exclusively by the pipes, he began to master the problem of uniform heating. Where in the past a shoe might emerge with the lower portion blistered and scorched and the upper portion barely changed, or vice versa, now even complex shapes started emerging evenly vulcanized with promising frequency. Just as important as the oven, Goodyear was getting a handle on the proper combination of rubber, additives, temperature, and time. The recipe that worked best consisted of twenty-five parts rubber to seven parts white lead to five parts sulfur. Lead, which Goodyear at first had used as a coloring agent, had proven useful in accelerating the vulcanization process. As for temperature, Goodyear found that objects could be vulcanized within a range of 245 to 300 degrees, but that 270 was optimum for most objects. Since the ovens had no external temperature controls to maintain a constant heat, Goodyear and his workers relied on a thermometer hung inside the box. They grew adept at controlling temperature by adding different amounts of coal to the fire. The time needed to complete vulcanization remained more an art than a science. It depended on the thickness and consistency of the material, but knowing when an object was "done" required the experienced eye of a chef who knows when to remove his soufflé from the oven.

Encouraging as Goodyear's progress was, it did not proceed quickly enough to please everyone. Henry Goodyear, a year or two younger than Charles, had a less inventive but more practical mind than did his older brother. Of all the Goodyear brothers, Henry had always been the most skeptical of Charles's rubber infatuation. Henry had joined Charles briefly in Philadelphia during the winter of 1834–35. Later, he had joined Charles at the Bank Street factory in New York City and then followed him to Staten Island. The bleak time of hunger and failure on Staten Island had convinced Henry that his brother was doomed, his failures contagious. He left Staten Island about the time Goodyear left for his visit to Roxbury. In 1838, he had briefly visited his brother in Roxbury, and Charles had asked him to assist in experiments. But while Nelson and Amasa Jr., along with Goodyear's parents, were frequent and lengthy parts of the household, Henry had stayed away: "I had seen so much suffering, in a pecuniary way, on his part and his family, and also of his own, that I refused to do so. I returned to New Haven, and in the fall I went South."

Henry had next seen his brother in Springfield and then in Woburn in 1841. Again, Charles asked for his help. Again, Henry refused to become involved. "He told me he had discovered a valuable improvement in baking and manufacturing India rubber, which was very valuable; that he had overcome great objections to the articles being affected by cold and heat. I declined stopping to assist him, as I had not much confidence in the improvements, from what little I saw of them at the time," he recalled. Finally, in the fall of 1842, Charles wore down Henry's resistance: "Finding that his experiments were more successful than I had anticipated, at his request I consented to stop and take charge of his factory, at Springfield." He would remain there until July 1844. But Henry quickly found his old frustrations with his brother's sloppy business habits resurfacing. Vulcanization, while improving continually, was still too unpredictable and unreliable for the mass marketing of goods: "In the fall and winter of 1842 we were unable to heat any thing, except small pieces, that were perfect; most of the goods were either ruined by being blistered or burnt in the heater, and were not considered fit for market." As Charles traveled restlessly from Springfield to Woburn to Lynn, Henry Goodyear was charged with the daily operation of the factory and the supervision of the workers. Yet Charles had final say on all matters, and he ruled out selling anything the factory produced. Stung by all of his past failures and the humiliation of rushing faulty goods to the market, Charles was determined to perfect the process before risking another debacle. But few people shared Charles's patience, his inexhaustible love for experimentation, and his odd, holy conception of rubber. It wasn't simply that Goodyear wanted to turn out the best product he could. That, Henry might have understood. But Charles seemed unconcerned with matters of commerce, and entirely unable to focus his attention on producing one thing and producing it well. To Henry, and to so many others, rubber was a commodity, a substance that might be worked to produce a payoff. Once again, a partner's desire to earn a living clashed with Charles Goodyear's dreaming.

Horace Cutler, a Springfield shoe manufacturer who occupied a building close to Goodyear's factory, was another who would join Goodyear, only to grow disillusioned. In the end, Cutler's defection cost Goodyear dearly. Cutler was initially swayed by Goodyear's passion for rubber when the two met in 1842. Impressed by samples of vulcanized fabric that Goodyear showed him, Cutler asked whether the material could be reliably produced. Goodyear conceded that a few kinks still needed to be worked out. But no matter. They were relatively minor impediments—he was on the verge of a breakthrough that would make him wealthy. Goodyear did not mention that he had been on the verge of a breakthrough for the past eight years. His optimism, ever trembling on the edge of self-delusion, remained his defining characteristic.

Cutler paid Goodyear three hundred dollars for the right to join him in manufacturing rubber overshoes. But as batches of finished shoes began emerging from the oven, some vulcanized, others blistered and useless, Cutler's patience wore thin. Pressed for cash, Cutler told Goodyear by the late summer of 1842 that he wanted out of the business and wanted his money back. Goodyear promised to return Cutler's money, which of course had already been spent. He also promised to give Cutler a job that would pay by the month. Cutler repeatedly asked him to come through on the promises, but Goodyear never did. The experience left Cutler bitter and angry.

Goodyear and Cutler had made several hundred pair of rubber shoes and boots. With the still uncertain methods of vulcanization, many shoes came out badly and had to be discarded. But the best among them, around ninety-five pair, Cutler consigned to one Horace H. Day. Day was a rubber manufacturer who ran a factory in New Brunswick, New Jersey, and a store in New York City. Day paid Cutler $26.75 for the shipment. It was a relatively minor transaction, thoroughly unremarkable in the great stream of commerce, except that it introduced Horace Day into Charles Goodyear's life—Day would remain a singularly terrible presence for the rest of it and, indeed, beyond.

Horace H. Day was a man so dastardly that in retrospect you almost have to like him—almost. He was a venal and hard-nosed competitor who spent much of his professional life launching or defending lawsuits. When not in court, he busied himself giving speeches or writing screeds to newspapers denouncing whoever opposed him. He was a pragmatic opportunist and yet he spent much of the second half of his life promoting the half-baked but popular quasi-religion of spiritualism, founded on the principle that one could commune with the dead. He made and lost fortunes. He was a cheat, an adulterer, and a liar. Horace Day drew his energy from fighting; he made many enemies, among them politicians, factory owners, and patent office managers. He was tall and physically powerful. He walked with his chest out and his full, long hair brushed straight back from his head, giving the impression that he was always on the charge. He liked pushing people around. And he reserved a special, curious dislike for Charles Goodyear, even though

Goodyear's perseverance made Day's fortune possible. To Horace Day, a covetous acquirer of money and material goods, Goodyear, with his overweening piety, was as alien as a specimen from some distant galaxy. He didn't trust a man who saw his business as a holy mission. He had nothing but contempt for Goodyear's sloppy business practices, for his poverty, for his genius. Horace Day made it his mission in life to suck Charles Goodyear dry.

Day was born in Great Barrington, Massachusetts, in the serene Berkshire Mountains, in 1813. When still a young boy, he was sent to live in New Brunswick, New Jersey, with an uncle, Samuel Day. New Brunswick was a grubby town of fewer than five thousand residents on the banks of the Raritan River. The streets were unpaved, and there was no gas or sewers. Municipal water came from wells. Sewage from the higher portions of the city frequently seeped into the wells in the lower portions, causing frequent outbreaks of typhoid. New Brunswick, home of Rutgers University, sits today on the fringes of the New York megalopolis, sixty miles from Manhattan on the New Jersey Turnpike. It is difficult to comprehend just how isolated those sixty miles made New Brunswick in the early nineteenth century. Roads and bridges connecting cities were nonexistent or hopelessly crude. The preferred method of travel to New York was not by train, horse, or stage, but by boat. The boat trip, down to the mouth of the Raritan and around Staten Island, took eighteen hours in good weather, three days in bad. New Brunswick by 1830 was just waking from long indolence and beginning to develop as a maritime center. The town supported a shipbuilding industry along the river, but the main commerce was in exports of agricultural products from the rich soil of the surrounding Raritan Valley. Long snakes of Conestoga wagons, as many as five hundred in a single day, filed down to the port to unload flour, pork, leather, and other local products. Ships that carried goods away from New Brunswick returned with cargo from wherever they had been. James Bishop, a young ship owner based in New Brunswick, learned of the seemingly endless supply of a new and remarkable substance tapped from trees growing along the banks of the Amazon River in the Pará region of Brazil. Soon, Bishop was importing large cargoes of raw rubber through the port at New Brunswick.

Among those taking note of the new miracle substance passing through town from South America was Horace Day, just coming of age in the early 1830s and casting about for something to do. With his uncle's assistance, he set up a small shop on Dennis Street near the river to make carriage tops and other goods. By the mid- to late-1830s, he was turning out carriage cloths coated with rubber, as well as shoes with rubber uppers attached to leather bottoms. He prospered initially, enough that he could afford to hire a young German immigrant named Christopher Meyer as a mechanic. Despite the general troubles plaguing the rubber industry, Day managed to stay afloat. In 1840, he opened a store in New York City, at Courtlandt Place, and a warehouse on Maiden Lane. But inevitably he suffered the same frustrations and setbacks familiar to everyone in the trade—when his goods failed in the heat or cold.

Sales were declining. Like everyone else in the rubber business, he desperately needed a break. Then he heard of Goodyear's work. The $26.75 he spent for the ninety-five pairs of crude rubber shoes from Goodyear's Springfield factory, purchased through Horace Cutler, were a bargain compared with the intelligence he hoped to glean from them. The shoes may have been crude, but Day was smart enough to realize the great leap forward they represented. Unlike every piece of rubber Horace Day had ever manufactured, these would not melt in the heat or crack in the cold.

Horace Day would never be content merely to buy rubber goods from Goodyear. He vowed to obtain the inventor's secrets. His efforts began in earnest on December 3, 1842, with a letter to Cutler, asking him to visit his New York warehouse at 45 Maiden Lane. Day instructed, "Bring along samples of every thing you have or can make. Should you have any Buffalo shoes ready made, bring one small box, and I can buy them possibly for retailing at our warehouse. I wish to know if you can cut the thread rubber which G. uses for suspenders." It's safe to say Horace Day had more on his mind than a small box of rubber-covered buffalo shoes. He added a postscript urging Cutler to keep the letter and the visit a secret.

Cutler, still angry at Goodyear, appeared as summoned at Day's New York warehouse two weeks later, on December 17.

"Did you bring the Buffalo shoes?" Day asked.

Cutler produced the box.

Day nodded approvingly. "What else did you bring? What can you show me that was made by Goodyear?"

Cutler handed over some samples of shirred fabrics—cloth made with embedded strips of rubber. "Do you know how Goodyear makes the threads that go into these fabrics?" Day asked.

Cutler nodded.

Cutler seemed nervous and reticent, so Day didn't press for details. Instead, he asked, "What would I have to pay you to come down to New Brunswick and work for me, and teach me Goodyear's method of compounding and heating India rubber? And making the shirred goods?"

Cutler considered for a few moments.

"I could do it for thirty dollars per month."

They shook hands on the deal. Before Cutler returned to Springfield, Day told him again to keep the plans secret. He also told Cutler that when he arrived in New Brunswick, he was not to tell anyone where he came from. "Just say you are from Massachusetts. Don't name the town."

In early January, Cutler arrived in New Brunswick, taking a room near Day's factory in a house owned by a widow named Mrs. Hall. Cutler soon took a dislike to New Brunswick, to Day's factory, and to Day himself. Sullen and homesick and uneasy with the underhanded dealings he was participating in, Cutler told Day he needed to go home immediately. Day offered a one-time payoff if Cutler would stay just long enough to divulge what he knew of Goodyear's methods. Cutler decided he could placate his guilty conscience for seventy-five dollars.

"Too much," Day snapped. "Look here, if you don't tell me, I'll just find out another way."

"How much will you pay me, then?" Cutler said.

"If you tell me the process, and if it works as you say, fifty dollars plus expenses for your trip here and back to Springfield," Day offered. Cutler agreed. Over the course of the next day or two, for fifty dollars, plus sixteen and change for boat fare and board at Widow Hall's, Horace Cutler systematically passed on to Horace Day secrets that had cost Charles Goodyear eight years of his life, his health, thousands of dollars, untold suffering, humiliation, ridicule, and poverty.

First, Cutler produced some small samples of a raw rubber-sulfur–lead compound that he'd brought from Goodyear's factory, asking Day where they could heat them. Day hustled Cutler into his boiler room, carefully locking the door behind them. Cutler slid out a pan of hot coals from under the boiler, fastened a piece of rubber to a stick, and started toasting the sample. When Cutler began to sweat in the thick, fetid air of the boiler room, Day took over the toasting. As he watched the rubber sample withstand the prolonged heat without melting into the fire, a smile, rendered devilish by the glowing coals in the dark room, formed on Day's face.

"It *will* do it, won't it!" he exclaimed.

Satisfied, Day ushered Cutler to the finishing room, where rubber goods were prepared for sale. Here, the shoemaker showed Day how Goodyear assembled his shirred goods, implanting strips of vulcanized rubber into cloth. As if building to a crescendo, Day next asked Cutler to tell him the precise ingredients.

Day was fascinated. "White lead, like painters use?"

"That's right."

"When you say sulfur, do you mean sulfur, or brimstone?"

"Sulfur."

"How much of each?"

Cutler gave him the formula. Day then asked Cutler to tell him about Goodyear's furnace. Cutler related the details about the cast-iron sides, the pipes to spread the heat evenly, the vertical reel for rotating the rubber samples, the thermometer to regulate temperature. Day took Cutler to an old foundry near his factory and asked if this would be a good place to build an oven. It would, Cutler said. He asked whether brick or cast iron was best. Cutler said Goodyear used cast iron, which held the heat better. With the interrogation complete, Day sat Cutler down in his office with a pen, paper, and ink, and demanded that Cutler put in writing all of the procedures and formulas he had just shared.

Having eviscerated his erstwhile partner, Cutler at last suffered an attack of conscience. Actually, it wasn't so much conscience as fear of possible litigation. Cutler remembered that Goodyear already owned Hayward's patent for combining sulfur with rubber, and had formally filed his intent with the patent office to protect his idea of heating rubber, sulfur, and white lead.

"If you manufacture goods under Goodyear's inventions, without a license, you'll have to take responsibility," Cutler said.

"Never mind," Day said impatiently. "I'll handle that thing."

❁ ❁ ❁

While Cutler skulked back to Springfield, Day set about in a fever trying to produce vulcanized rubber objects. Years later, when fighting Goodyear in court, Day would claim that he had actually invented vulcanization before Goodyear did. But for now, all he wanted to do was reproduce the magic that had been as tantalizingly close as the hot gob of rubber at the end of Cutler's stick.

Try as he might, though, Day could not get the procedure to work on anything but the smallest scraps of rubber cloth. Of course, Day, despite the enormous head start provided by Cutler, was merely encountering the same difficulties of temperature, mixture, thickness, and so forth that had dogged Goodyear for so long. But Day lacked Goodyear's patience. And Day had much more of a mind for profits. One can only imagine the fury and frustration of the excitable Horace Day as he watched yet another batch of products rendered blistered and worthless. Instead of feeling contrite for having stolen Goodyear's hard-won intelligence, he was filled with rage that he could not reproduce the process. As 1843 wore on, Day wrote exasperated letters to Cutler, demanding to know why everything he tried came out blistered and wrecked. But Cutler, for all his knowledge, was still essentially an outsider to Goodyear's world. He had no more knowledge to share with Day than what he had already given.

So Day went to the source. In the summer of 1843, in a typical act of hubris, Day traveled to Springfield to tour Goodyear's factory. When he arrived, Henry Goodyear met him at the door with polite reservation.

"I wonder if I could have the privilege of looking at the works?" Day said obsequiously.

"I'm sorry, it's against the rules of the factory," Henry replied.

"I am very well acquainted with Charles," Day lied. "If he knew I was here he would certainly let me in to examine the works."

"Really," Henry said. "Well, he's upstairs right now. I will get him."

A few moments later Goodyear appeared at the door. There's no record of the conversation that transpired. Too bad—it must have been priceless. Goodyear politely but firmly refused to allow Day into the building. As he watched Day walk away from the factory door, Goodyear, unaware of Day's connection with Cutler, probably assumed he had rid himself of an annoying but minor interloper.

For the time being, Goodyear had more pressing matters to worry about than Horace Day. The Riders, his great and generous friends from New York, suffered a major financial setback. With the failure of the Riders, Goodyear once again found himself short on cash and overextended. Every cent had gone toward equipment and material. One day the local sheriff, a man named Foster, arrived at the factory. For Goodyear it was a familiar story with a new location. For nonpayment of some debt, Foster escorted Goodyear to the Springfield jail. As prison spells went, this experience, his last in the United States, was better than most. Foster declined to lock up the inventor, instead giving him free run of the jailhouse. Goodyear had recently ordered a new suit of clothes. When the manager of Palmer & Clark, the tailor shop, heard of Goodyear's arrest, he might well have stopped delivery. Instead, he kindly directed one of his young clerks to deliver the suit to Goodyear at the jail, with the request that Goodyear pay whenever he was able. The clerk, whose name is now unfortunately lost, stopped home on his way to the jail. His mother asked him to remove the head from a barrel of apples that had just arrived. He did, and then, on his way out the door, plucked a large, juicy apple as a gift to the convict. When he arrived at the jail, he found him sitting comfortably in Sheriff Foster's office, reading a book. He presented the suit to Goodyear, and then the apple, which Goodyear gratefully accepted. Goodyear, who never forgot a kindness done to him, was deeply moved by the gesture. Nine years later, Goodyear would spot the clerk by chance in the office of an American businessman in Paris, where the clerk was then working. Goodyear, having become a celebrated figure, recognized the man, introduced himself, and recalled the apple incident in every detail. He then invited the astounded clerk to ride with the inventor and Emperor Louis-Napoléon, nephew of Bonaparte, to the Bois de Boulogne that afternoon, in the emperor's private carriage, drawn by four magnificent horses.

Goodyear's continuing poverty in Springfield only partially explains what hindsight clearly reveals as a series of colossal strategic blunders that together would cause Goodyear years of trouble, rob him of a probable fortune, and for a time at least threaten his claim as the inventor of vulcanized rubber. The first mistake was his ill-advised delay in taking out a vulcanization patent, either in the United States or in Europe. Certainly, money had something to do with it—a man still getting thrown into jail for debt could hardly afford to file for U.S. patents, let alone overseas ones. And yet he might have done so a year or two earlier, in the first flush of cash from the Riders. Indeed, it was the Riders who had insisted that he at least file his letter of intent with the U.S. patent office. More likely, the delay owed itself to the same reason that Goodyear refused to market vulcanized goods—he wanted to perfect the process first. He had at last grown wary of promoting ideas only to see them collapse.

The next mistake, ultimately far more damaging than his patent delay, involved his efforts to attract foreign (specifically, British) investment capital. In one sense, it seemed a logical enough move. Great Britain, after all, was the world's industrial engine, next to which the United States was an eager but scrawny child. Moreover, Goodyear knew he was running out of potential American investors in rubber, having already made his way through Connecticut, Massachusetts, and New York.

Goodyear turned to Stephen Moulton, the British-born broker whom he had met through the Riders. Moulton and Goodyear remained friends even after the Riders' financial failure curtailed Goodyear's partnership with them. Moulton had been planning a trip home to England anyway and agreed to help. Despite his best intentions on Goodyear's behalf, Stephen Moulton might as well have been delivering scraps of fresh meat to a lion's den.

9

TO THE VICTOR

The steamer *British Queen* slipped from its New York berth in August 1842 with a few pieces of cargo too small to appear on any manifest. They were small enough, in fact, to fit neatly in Stephen Moulton's breast pocket. But these modest strips of vulcanized rubber were undoubtedly the most important cargo on board. They would at last bring two worlds—that of Charles Goodyear's ragged quest and of Thomas Hancock's industrial powerhouse—together.

Moulton was determined to keep Goodyear's name a secret, but that hardly mattered. Goodyear was still an obscure figure on the fringes of the scientific community in his own country; in England he was a nonentity. There was another secret, of course, one infinitely more important. But Goodyear believed the details of his vulcanization process to be safe. Moulton seemed the perfect conduit. He was honest, fair, and both sympathetic to Goodyear's plight and a believer in the ultimate possibilities of his invention. He was familiar with business practices and customs on both sides of the Atlantic. And he had just the right level of knowledge—he could speak the basic language of rubber but had little or no idea of what specifically Goodyear's process entailed. He could not divulge the secret even if he wanted to.

Goodyear asked Moulton to show the samples to important rubber manufacturers in Great Britain. Fifty thousand pounds—his asking price for the process—would enable Goodyear to continue his experiments for the foreseeable future. Moulton ultimately showed the samples to several people during his trip, but the only individuals who really mattered were the officers of Charles Macintosh & Company. Macintosh himself, approaching seventy-six and with less than a year to live, by now had little practical involvement with the company that bore his name. George Macintosh, his intelligent but rather ineffectual son, exerted little authority despite his status as partner. The Birleys, Charles Macintosh's original Manchester partners, oversaw the Manchester plant while Hancock provided overall direction for the company from London.

Moulton stopped first in London. He did not know Hancock personally, but he and Hancock did have a mutual acquaintance named William Brockedon. Moulton asked Brockedon to deliver the samples to Hancock and put in a good word on behalf of his client. Brockedon examined the samples and "was much struck with their properties of resistance to heat or cold," as Moulton later recalled. He agreed to show them to Hancock. If his portrait, a slide of which can be found at the Science Museum in London, is any indication, Brockedon was something of a fop. In the picture he wears a high collar and flowing tails, his face framed by long sideburns. He holds an elaborate quill pen in his right hand. His contemplative expression tells us the open book resting on a pedestal is about to be filled with deathless prose. For all that, Brockedon was a comparatively minor figure in the rubber industry. His great obsession was to develop a new bottle stopper to replace cork. His one lasting contribution—giving vulcanization its name—was at least two years away.

Moulton did not linger in London awaiting an answer. He continued on to Manchester, where he showed additional samples to the Birleys. They, too, were impressed. But when Moulton quoted the price, they balked. They simply could not part with £50,000 on the basis of a few scraps. Yes, the samples were impressive. But how did they know the process could be reliably repeated? How could they estimate the cost of manufacture without knowing what it entailed? The Birleys told Moulton that his inventor friend would do well to patent the process. Then he could freely share the details under protection, and Charles Macintosh & Company could make a more informed decision.

As Moulton prepared to leave, the Birleys asked if they could keep the samples. Moulton agreed. An ominous exchange followed. According to Moulton, one of the Birleys asked, "But what if we should discover the secret?" In that case, Moulton replied, any such knowledge must remain in the strictest confidence. He added that the inventor "must for this contingency trust entirely to circumstance and the honor" of Charles Macintosh & Company. By all appearances, the Birleys were forthright and honest during this meeting. Their objections to buying the process blind were reasonable, their advice on taking a patent sound. Had Goodyear done so in the first place, he never would have suffered the anguish that was to follow. Moulton must have left Manchester reassured.

If William Brockedon had been impressed by Stephen Moulton's rubber samples, Thomas Hancock could scarcely believe his eyes. Hancock, the acknowledged authority on rubber in the most powerful industrial nation on earth, had long ago implicitly declared, by his inaction, that rubber would always be susceptible to heat and cold. There had been no significant work on the problem in England for years. These samples by their very existence rebuked Hancock's decision to concentrate on the safer course, on incremental improvements.

All of his contributions paled to insignificance next to the leap forward represented by the pieces in his hands. What was the masticator next to this?

But his reaction wasn't just emotional. Whoever had the rights to this new material would control the rubber market. That much was clear. If the owner were someone other than Thomas Hancock, Charles Macintosh & Company might well be finished. There was an alternative, of course: take the honorable course, seek out this unknown American, and offer him a deal. Goodyear would have leaped at almost any proposition, considering his chronic poverty and debts. Yet if Hancock felt any compunction or desire to find the inventor, the urge died quickly.

Certainly, Hancock was not legally bound to offer Goodyear anything. Nor was it illegal to use the samples to deduce the secrets for himself. Hancock, after all, had not stolen the samples, or hired a single spy. The samples fell into his lap. The American had made a serious mistake by not taking out a British patent before sending the material over. The window was wide open—Hancock could not restrain himself from climbing through. Through long and diligent effort he had essentially built an industry from scratch. His reputation was hard-earned and sterling. He was the Father of the Rubber Industry. This single act would affix to his name, despite all of his genuine contributions, the indelible mark of a sneak.

But for now, he set to work in a fever. Given the challenge before him—unlocking the secrets to an invention he had until now believed impossible—Hancock might well have given up before he even started, except that Goodyear had unwittingly provided some tantalizing clues. The darkened edges of the samples suggested that heat was an essential component. But there was more.

To this day, one of the principal challenges in vulcanization is applying the appropriate heat and cooking time relative to the sulfur content, to allow the sulfur to bond fully with the rubber. Otherwise, orphan sulfur particles have a disconcerting tendency to work their way slowly to the surface of the material, creating a yellowish rash known as a sulfur bloom. By 1842, Goodyear was an expert at heating rubber and sulfur. Even so, his methods and measurements were necessarily inexact. With the high levels of sulfur he used and the vagaries of nineteenth-century ovens, there was simply no way to avoid imperfections. Sometime during Moulton's voyage, sulfur, the very substance that had helped Goodyear make his great discovery, betrayed him by rising to the surface like a clue-laden body from the depths of a lake. By the time the samples came to Hancock, at least one contained an unmistakable bloom. Heat and sulfur. Sulfur and heat. Aha!

Hancock nevertheless had his work cut out for him. Recall that Horace Day had received from Horace Cutler the ingredients for vulcanized rubber, their precise proportions, and a live demonstration, and was *still* unable to reproduce it. But Horace Day was no Thomas Hancock. Hancock had forgotten more on the subject of rubber than Horace Day would ever know. Hancock was brilliant, well-organized, and methodical. And now he was motivated.

Hancock's experiments were too sensitive for the London factory. As was his habit whenever his work demanded secrecy, he confined his work to the attic laboratory of his home in Stoke Newington. After putting in a full day tending to the affairs of Macintosh & Company, Hancock returned home, ate a hasty dinner, then climbed a ladder to an attic retreat. It was a small, workmanlike room with light and circulation provided by a pair of large windows. There were few comforts of any sort. Work tables supported heavy tools such as vise grips and clamps. Over the fireplace were six large shelves neatly stocked with more than a hundred bottles of powdered and liquid chemicals. Hancock's desk was a plain wooden table near a window, with a straight-backed wooden chair.

No servants were permitted in the room. Hancock lit his own fire for warmth and cleaned up after himself. Often he worked until midnight, then got up the next day and repeated the process. It was a schedule that might quickly have exhausted a much younger man, let alone one in his late fifties.

His progress for the first few months was frustratingly slow. Although Hancock, thanks to Goodyear, had a rough idea of the ingredients for vulcanized rubber, he had no idea how to mix them. Did they undergo the change while in a liquid state or a solid one? Hancock softened rubber in his masticators, then mixed it with turpentine and sulfur to create a thin liquid, which he spread on canvas in hopes it would dry into vulcanized rubber. No luck. Then he tried adding more rubber and sulfur, less solvent, to create a thick dough. This, too, failed to yield anything promising. Sometimes, he heated samples by laying them on a metal plate which he held over a lamp. But the heat was too meager to produce much effect. As the winter of 1842–43 deepened, the pile of discarded rubber scraps grew until it contained thousands of pieces. Hancock separated these into two piles, one vastly larger than the other. In the large pile lay heaps of absolute rejects. In the smaller pile, the few scraps that seemed to hold some promise. The expense of these experiments, in both time and materials, must have reminded him sorely of the "rather severe loss" that had made him give up the quest in the first place. But now there was an imperative. Rubber *could* be made to change. And somebody in America had already done it.

While Hancock toiled in private, Goodyear whiled away months conducting further experiments, confident that his secret was safe. After returning to New York in November, Moulton corresponded on Goodyear's behalf with both the Birleys and Brockedon over the next several months. In April 1843, Moulton visited England again. Brockedon told Moulton that the American inventor should come to England to meet the principals of Charles Macintosh & Company. According to Moulton, Brockedon made a point of saying he thought it "next to impossible" that anyone could figure out the secret process based on the samples. Whether Brockedon was being genuine, or simply buying time for Hancock by putting Moulton at ease, can never be known for certain. There are good reasons to suspect the latter. In any case, the correspondence and visits ate up precious time, and Thomas Hancock at last was beginning to make progress.

Hancock had started dissolving sulfur in turpentine over a hot stove. When the mixture boiled, he added rubber, which quickly dissolved. The solution, when spread on strips of cloth, seemed somewhat more resistant than natural rubber to heat and cold. Although the resulting material was weak and the results inconclusive, Hancock was at last nibbling at the edges of vulcanization, looking for the appropriate combination of sulfur, heat, and rubber. As his experiments progressed into the heat of the summer, Hancock hurried downstairs each morning to meet the passing ice wagon, buying blocks to test samples that night for resistance to the cold. "My experiments had now become very interesting," he recalled in his memoirs. "I had certainly produced in some of my scraps, or portions of some of them, that condition of rubber which I afterwards called the change." On July 17, Charles Macintosh was struck with severe diarrhea. He grew weaker over the next few days and, sensing his death might be imminent, sent for Hancock. Hancock set aside his experiments and hurried to Scotland, but Macintosh died, on July 25, at seventy-six, before Hancock could get there. "I have never met with any person for whom I entertained a greater esteem," Hancock said of his old partner.

By the fall of 1843, Hancock had been working on vulcanization for a solid year. He had his ingredients and a general idea of what to do with them, but he lacked a reliable formula. He decided to apply for a provisional patent. This represented a gamble on Hancock's part. Under British law, he had six months to announce the specifications for his invention—specifications that as yet did not exist. If he could not produce an exact formula at the end of that time, he would lose the patent. As it turned out, Hancock's timing was perfect, much to Goodyear's detriment. Hancock received his provisional patent on November 21. Two months later, Goodyear's own application arrived at the British patent office.

With just six months to succeed or lose out, Hancock redoubled his efforts. The major difference between Goodyear's method and the one devised by Hancock from Goodyear's samples was that Hancock knew nothing of the use of white lead to accelerate the process. Hancock used only sulfur and rubber. If simply dipping rubber in a molten sulfur bath produced promising but flawed results, Hancock reasoned, perhaps he should be mixing rubber with sulfur, then dipping the mixture in the molten sulfur. Here was his breakthrough. Rubber samples began to emerge from his laboratory fully cured.

Exactly six months after applying for the provisional patent, Hancock filed his formal application with the English patent office. If Goodyear ever read the patent, it must have torn at his heart, for in a few dry, matter-of-fact words Hancock usurped the great accomplishment of Goodyear's life: "The nature of my improvement or improvements in the preparation or manufacture of caoutchouc in combination with other substances, consists in diminishing or obviating their clammy adhesiveness, and also in diminishing or entirely preventing their tendency to stiffen and harden by cold, and to become softened or decomposed by heat, grease, or oil."

Hancock further explained: "I diminish or obviate these effects by intimately blending sulphur with caoutchouc in various ways, but the following I find to answer the desired purpose:—I melt in an iron vessel a quantity of sulphur, at a temperature ranging from about 240° to 250° Fahrenheit, and immerse it in the caoutchouc, previously rolled into rough sheets, or cut to any convenient form or size, and allow it to remain until the sulphur has quite penetrated the caoutchouc . . ." Sometimes Hancock sliced the rubber with a wet knife to allow the sulfur to penetrate fully. Once the piece was saturated (it turned yellowish through and through), he dipped the item in a hot bath of molten sulfur. Hancock also mixed sulfur and rubber with solvents, especially when making rubber-covered cloth. These pieces he submitted to an oven.

Hancock's defenders would later make much of the fact that Hancock used no white lead, as though Hancock's recipe of just rubber and sulfur represented a more "pure" or "fundamental" discovery. But the process of sulfur and heat cross-linking and strengthening rubber macromolecules was essentially identical.

Like all developments, this new process demanded a name. Hancock at first referred to the process as, simply, "the change." Goodyear called the results "metallic gum-elastic" or "fire-proof gum." Hancock's description was too broad, Goodyear's too cumbersome. The enduring name came from an unlikely source—William Brockedon, the man who had shown the samples to Hancock. Brockedon, delighted with Hancock's results, suggested "vulcanization" after Vulcan, the Roman god of fire. Vulcanization had a nice ring to it—scientific and serious sounding, yet with a certain flair. Hancock liked it and began using the term in Macintosh & Company advertisements. Within a few years people on both sides of the Atlantic were referring to the process as vulcanization, including Goodyear, who refers to "vulcanization" throughout his memoirs. Hancock's corporate descendants, James Lyne Hancock Ltd., reissued Hancock's memoirs in 1920 with a pugnacious defense of their ancestor. The naming of vulcanization by Brockedon and Hancock, the directors insisted, offers "striking proof of Hancock's undoubted right to be recognized as the discoverer of Vulcanization." As if the first person to name the big ball in the night sky could lay claim to having discovered the moon.

One of the few to object strongly to the term was Stephen Moulton, Goodyear's friend. Moulton, destined to become an important British rubber manufacturer in his own right, would remain fiercely loyal to Goodyear and critical of his fellow countryman Hancock. He chastised an American associate in 1848, "How could you possibly adopt the title of Macintosh [& Co.] and call your rubber Vulcan-ized? I wish you could teach me *the meaning* of the word. It has none whatever, and your effect in using it appears to me an exceedingly foolish one."

Although Goodyear never blamed Moulton, Moulton always regretted his unwitting role in depriving his American friend of the English patent. And he would hold a particular grudge against Brockedon, whom he suspected of playing more than a passing part. When the two met after Hancock's patent victory, Moulton openly accused Brockedon and Hancock of stealing Goodyear's invention. According to Moulton, Brockedon replied with a mischievous grin, "That is very good!" and then boasted that it was he who first noticed and alerted Hancock to the sulfur bloom on Goodyear's samples.

Later, under oath, Brockedon would deny everything—that he noticed sulfur, mentioned it to Hancock, or boasted to Moulton. Whether Brockedon received a direct quid pro quo can never be known for sure, but there's no denying that he made out handsomely from the association with Hancock. In 1845, when original partner Hugh Hornby Birley died and less than two years after the death of Charles Macintosh, Charles Macintosh & Company formally reorganized its partnership. George Macintosh, son of the founder, though not yet fifty-five years old, was eased out. The new partnership consisted of Thomas Hancock, the remaining Birleys—and William Brockedon.

❁ ❁ ❁

Thwarted in his belated quest for the great prize, the English patent, Goodyear had to make do with the less valuable U.S. patent, which he had applied for in January 1844 and which was approved June 15, 1844.

Goodyear was deeply hurt by Hancock's actions, less by the financial loss than because it had clouded the issue of who had made the breakthrough. Though he shared with close friends his contempt for Hancock, Goodyear remained publicly circumspect. Reflecting on the episode a decade later in his book, *Gum-Elastic and Its Varieties*, Goodyear chose his words carefully: "The writer has intentionally omitted all mention of the English patent for the vulcanization of caoutchouc, enrolled by Thomas Hancock on the 30th May, 1844, and also from expressing any views or opinions in relation to the circumstances under which this patent was taken out, lest he might possibly do some injustice to Mr. Hancock or his partners, Messrs. Macintosh and Co., the owners of the English patent. The invention patented by Mr. Hancock is the same as that so fully described in this volume as the heating or vulcanizing process discovered by the writer in 1839."

Goodyear had spent many years being called a fool for obsessing on the vulcanization problem long after sensible men had given up. Now he found that he had persevered only to see credit for the breakthrough slipping through his grasp. And his fight was just starting, not only with Hancock but with a whole world that suddenly wanted money or credit, or both, for a development that would not have existed without Goodyear's obsession.

In addition to his U.S. patent, Goodyear in April 1844 was awarded a French patent for vulcanized rubber. But any satisfaction on this front would be short-lived. French patent laws demanded that any invention be brand-new to the country; that is, an inventor could not patent any invention if he had already sold goods based on the idea in France. Goodyear had shipped a few pairs of vulcanized shoes to France for sale before 1844. When a competitor brought this to light, the government stripped Goodyear of his patent and essentially declared vulcanization public property in France. It was the industrial equivalent of Jim Thorpe's losing his Olympic medals because he had once earned a few dollars playing semi-pro baseball.

Even with the severe disappointments over the English and French patents, Goodyear's U.S. patent should have been enough to earn him a fortune. Rubber, suddenly, was hot again. Thanks to Goodyear's breakthrough, investors and would-be manufacturers were as eager to reenter the rubber business as they had been to desert it just a decade earlier. Vulcanization made possible the realization of all the fantastic predictions of that earlier time—now waterproof clothing, shoes, hats, carriage cloths, life preservers, and so forth could be made to truly live up to their billing. If manufacturers first concentrated on the products where rubber had been tried before, new and more exciting applications quickly arose. With the country growing rapidly and the mobility of its population exploding, two new avenues—industrial machinery and transportation—opened up. The automobile with its inflatable tire was still decades off, but vulcanized rubber could provide springs to soften the ride for railroad and carriage passengers. Vulcanized rubber would not disintegrate; it could be used as packing for steam engines, absorbing high impact without seeming to wear. As tough and reliable as unvulcanized rubber had been weak and fickle, this new product could be used for fire hoses and conveyer belts, and to seal plumbing joints. By licensing alone, Goodyear should have become a wealthy man, just as he had always assured friends and detractors alike that he would one day be.

That he failed to do so was the result of his own lack of understanding of money, his financial irresponsibility, and the concerted efforts of jackals such as Horace Day. Charles Goodyear was not built for making money. He spoke of one day having "comforts" and of being able to provide his long-suffering family with them, but always in the most abstract and dreamy way. Like many creative people, he lacked the essential concept of a link between financial rewards and action. Much as he might have wanted money, he had not the faintest idea of how to earn it. He could not bring himself to develop a passion for cash on a par with his passion for experimenting. Money simply annoyed him. The ship that Goodyear always vaguely believed was sure to come in for him was destined to remain a ghost ship on the horizon forever. Goodyear himself ultimately became resigned to this. He wrote, "The history of inventions as well as authors, with few exceptions, proves that whoever attempts by inventions to improve the conditions of others, usually impairs his own, except so far as he may add to his happiness, from the satisfaction of having done good to others."

Had he a stronger head for finance or a thirst for money, Goodyear would have concentrated his energies on a few reliable products that he could turn out over and over again, efficiently. But he was no incipient Henry Ford. Goodyear's tendency was to do the opposite, to slough off the boring but potentially remunerative products on other people. He could not abide the drudgery of repeatedly making something he already knew how to make. And so other people were destined to make their fortunes on Goodyear's idea.

William DeForest, Goodyear's brother-in-law and one-time tutor in Naugatuck, had always harbored a soft spot for Charles. He had taken a sort of paternalistic interest in him, sending money from time to time and lecturing him to give up his quest and pay more heed to the needs of his family. DeForest must have been shocked when his harmless lunatic of a brother-in-law, who had seemed to be frittering away his life and health chasing phantoms, turned out to be right. DeForest, a successful woolen manufacturer in Naugatuck, quickly transformed from tolerant patron into one of the first major rubber manufacturers in the United States. The company he started in Naugatuck would grow up to become the progenitor for the United States Rubber Company, which later became Uniroyal.

Emmet A. Saunders, a Naugatuck native and a key employee in the Naugatuck rubber works, offered perceptive contrast between these two men, in a quote unearthed by Naugatuck historian Constance Green:

> Here was DeForest, a man of immense energy, fine bodily presence, not much education, but with a tremendous faculty for trading and swapping. He was my "beau ideal" of a "fine old gentleman." Large frame, fairly pleasant face (red, red, very red), buff waistcoat, tail coat and brass buttons and tall hat, usually brown with long furry glistening nap, very shiny. He always spoke of Goodyear with a wholly loving, but half contemptuous accent, as if he were a dearly beloved "enfant terrible" that should not be held responsible for anything except to be his own blessed self.
>
> And here was Goodyear, a dreamer, reaching for the moon, careless about such small and entirely material things as food, shelter and clothing for himself and family. He would be dependent for these upon any friend, he would give anything he had if they wanted it more than he did just then, and he would take from anyone, borrow from anyone, not because he was a mendicant or "dead beat" but because these things were of little importance and sometime "when his ship came in" he would fix it all, immaterial and temporary. DeForest fed and housed him and his family when they had no other refuge. He was not the man to keep account and store

up items—but when Goodyear had anything that could be sold or traded because of DeForest's competence and Goodyear's incompetence, DeForest would take hold and help.

Now, DeForest pulled together a group of investors to form the Naugatuck India Rubber Company. The company paid Goodyear fifty thousand dollars, in exchange for his allowing the company unlimited (though not exclusive) patent rights for the present and future. Henry Goodyear, who had managed the Springfield plant, moved down to take over day-to-day operation of the Naugatuck plant. This was the end of the Springfield phase of Charles Goodyear's career. The plant was set up in an old button factory, a nice completion of a circle, of sorts, considering that Amasa Goodyear had come to Naugatuck nearly forty years earlier to start a button factory. Naugatuck in 1844 was still a sleepy, out-of-the-way community with just three hundred families, nearly all of them of English Yankee stock. By the end of the century, thanks largely to the rubber industry, Naugatuck was a bustling city of more than six thousand residents, nearly half of them Catholics. The lure of manufacturing jobs attracted waves of Irish, Germans, Swedes, and Poles. Long before the rise of Akron, Ohio, Naugatuck would develop into the nation's rubber capital.

Success of the Naugatuck venture was immediate and dramatic. There were actually two companies formed, the Goodyear Metallic Rubber Shoe Company and the Naugatuck India Rubber Company. The original capitalization of the two companies together was $70,000. Within a few months of starting operations, the mills were valued at $120,000.

Neither Charles Goodyear nor any of his descendants would grow rich from the success of the United States Rubber Company, since Goodyear had signed away his rights for a flat fee. It may seem that DeForest and others were taking advantage of him. But the deal had several key advantages for a man of Goodyear's temperament, disposition, and financial state. For one thing, by signing away the rights, he prevented past creditors (and there were plenty) from attaching their claims to the patent itself and disrupting any attempts to manufacture rubber goods. And Goodyear, as always, needed the quick cash more than he needed the possibility of future riches. In addition, Goodyear could earn money from auxiliary licenses stemming from his rubber patent. Under the deal, the Naugatuck Company had first refusal for any proposed product. But if the company decided it could not or did not wish to handle the product, Goodyear was free to sell the license to a third party. Finally, and this perhaps had the greatest appeal for Goodyear, the Naugatuck directors offered Goodyear the space and freedom to experiment further with rubber.

For Goodyear and his family, these were, finally, comfortable times. Though Goodyear never managed to fully extricate himself from years of debts, and although they would never enjoy life on a scale of luxury commensurate with the invention he had given the world, neither were they digging half-grown potatoes and facing constant eviction. Goodyear moved his family to New Haven. Connecticut must have been a sweet homecoming for the itinerant family, especially for Clarissa. They moved to comfortable quarters at 2 St. John Place, a building near the center of town, overlooking the green where his ancestor Stephen Goodyear once lived. Goodyear would spend seven years in New Haven—his longest stretch in any one city as an adult. But even in this relatively settled state, the restless inventor moved around. City records show his moving from 2 St. John to a home at 48 Olive Street, in an old residential section to the east of the green, in 1849 or 1850. In 1851 he moved his family to the Tontine, a large, prominent, three-story hotel near the green, where his brother Nelson had been boarding since 1846. Even with his association with the Naugatuck Company, Goodyear operated a factory inside the Union Works, amid a cluster of manufacturers near the waterfront.

He needed his own space by then, because the Naugatuck managers soon found Goodyear and his space-consuming, pie-in-the-sky experiments disruptive.

A man who saw rubber as proof of God's greatness, and himself as the messenger, simply could not confine himself to shoes and carriage springs. Not all of his ideas about rubber's uses were as brilliant as vulcanization had been. In the mid-1840s he grew enamored of the idea of rubber sails billowing on the masts of the world's great ships. Ship sails represented "perhaps the most extensive and important of all the applications of gum-elastic." Rubber sails would never freeze in the harsh Atlantic winters. Since they were airtight where canvas was not, they could provide more power from the available wind than could comparatively porous canvas. Their greater strength would make them less susceptible to tearing. With nary an order in hand, Goodyear set about designing and making large sails of duck cloth coated with rubber. They were inordinately expensive and time-consuming to make, and a particular technological challenge because of the immensity of the sheets. The shipping community, rubber licensees, and the public showed little interest. To an undaunted Goodyear, this was yet another example of the peculiar human tendency to resist great advancements—a phenomenon he knew only too well.

Charles W. Popham, the captain of the *Stephen Whitney*, a packet ship running the New York to Liverpool route, eventually did fix a Goodyear topsail to the masts, where it stayed for six Atlantic crossings, including fierce winter gales. He was favorably impressed. He reported to Goodyear that the sail was durable, remained pliable in the cold, and resisted mildew. A few months later, one of those Atlantic gales would send Popham and the *Stephen Whitney* to the bottom of the sea—the rubber sail, Goodyear noted with some relief, had been removed and thus could not have contributed to the disaster. But rubber sails would never find so enthusiastic an admirer as Popham. They were heavier than plain canvas and more cumbersome to store and handle. At any rate, the approaching age of steamships rendered this particular Goodyear dream moot.

While Goodyear was thus engaged, licensees such as Jonathan Trotter of Brooklyn turned out less fanciful but more profitable items such as canteens, knapsacks, ponchos, and shatterproof bottles. A partnership including John Haskins, Goodyear's old friend from Roxbury, made elastic bands. The Rider brothers, his benefactors from New York, produced mailbags (of the nonmelting variety).

Though he signed many license agreements, Goodyear rarely gave enough thought to working out advantageous deals. He owed so much on past debts that often he was in no position to bargain. He granted more than a few licenses simply as payment for old debts. Others, to whom Goodyear did not owe money, recognized an easy mark when they saw one, and made a habit of ripping him off.

The last time Goodyear had encountered Horace Day was outside the Springfield factory, where Day tried unsuccessfully to wheedle his way in. When Goodyear and his brother sent Day packing, Day decided on another tack. He would apply for his own patent, based solely on the information he had received from Cutler. Day, changing a few cursory details, simply rewrote Goodyear's process and submitted it as his own. On June 15, 1844, patent commissioner H. L. Ellsworth sent this terse reply: "Your application for Letters Patent, for an alleged improvement in preparing India rubber, has been examined, and rejected for want of novelty. An application for a patent for a similar process to yours was made on January last, by Charles Goodyear, of New York. For forms of withdrawal or appeal, see enclosed letter."

By now, Day was already establishing himself as a first-rate cad, in his private as well as his public life. In the fall of 1843, he was thirty years old and growing weary of life in the provincial backwater of New Brunswick. He began spending more and more time in New York City, where he had a retail store and warehouse. He and his wife, Sarah Day, had been married for six years, since 1838. Their son, Nicholas, was five years old, and Sarah was pregnant with their second child. But Day had fallen in love with his cousin, Catharine Day. On October 12, the day his daughter, Helen, was born, Day deserted the family for New York and took up with Catharine. Divorce papers filed by Sarah in April 1844 accuse Day of "wickedly disregarding the solemnity of his vows and the sanctity of the marriage state" and having "committed adultery with Catharine Day and divers other persons . . . unknown." In September, the court granted the divorce and ordered Day to pay alimony and child support. Day subsequently married Catharine Day.

Having failed to snag his illicit patent based on Goodyear's ideas, Day decided on the next best approach—poaching on one of Goodyear's most valuable licenses. Shirred goods were the most immediately popular and profitable items resulting from vulcanized rubber. In terms of the technology and the amount of rubber used, shirred goods represent a particularly modest beginning for the modern rubber industry, but it makes sense that they worked so well so quickly. The vulcanization was simple because the rubber pieces were flat strips. They revolutionized clothing. Shirred, or "corrugated," cloth involved long, slender strips of rubber being glued between two pieces of cloth. The rubber strips would be stretched tight before being glued and pressed into place. A machine would hold the strips in place until they dried. When loosened, the strips contracted, shirring the cloth. It was the perfect way to make snug-fitting caps, cuffs, stockings, gloves, and other items.

Shortly after Goodyear obtained his U.S. patent, a would-be rubber manufacturer named David L. Suydam paid Goodyear fifteen thousand dollars for exclusive rights to manufacture shirred goods. No sooner had Suydam begun manufacturing his products than Horace Day introduced his own line of shirred goods. Day manufactured hastily, with little attention to quality. The cheapness of his methods made for an inferior product, but they were less expensive to make, and Day was able to flood the market with inexpensive alternatives. Overwhelmed, Suydam had little heart for a fight. He asked Goodyear to rescind the contract, saying he wanted out of the business. Goodyear agreed, and then struck back at Day.

10

STRUGGLES OF ANOTHER SORT

For Goodyear and Hancock, the decade from the mid-1840s to the mid-1850s comprised the age of litigation. Vulcanized rubber had emerged from the obscurity of Goodyear's solitary struggles and the secrecy of Hancock's locked attic. Everyone wanted a piece.

Goodyear filed suits against Horace Day in 1844 and again in 1845, on behalf of himself and David Suydam, charging that Day infringed on Goodyear's patent and Suydam's license by flooding the market with inferior goods. Day rarely dallied on the defensive, preferring frontal assaults on his enemies. Ignoring the charges against him, he launched a countercharge that he would repeat for the next twenty years, whenever it suited him: Goodyear was a fraud. Day had no evidence to support this claim, but he was loud. And the shirred goods case was merely round one.

When Goodyear received his U.S. patent and, with it, some publicity and money, Nathaniel Hayward began to feel left out. Goodyear had always acknowledged Hayward's role in leading him to sulfur. It was Goodyear, after all, who had insisted that Hayward take out a patent for his "solarization" technique. But credit due to Hayward ended there. Hayward never pursued the process beyond laying rubber goods out in the sun.

In September 1844, when Hayward was feeling particularly resentful, he wrote a letter to Day. "I understand that Mr. Charles Goodyear has commenced a lawsuit against you for infringement. If so, please write me, for I am the inventor of the heated gum, and the improved gum, which I can give any information you may want on your case." Hayward and Day exchanged several letters, Hayward working his way into a slow boil of righteous indignation over having been cheated, Day fanning the flames. Day harped particularly on Goodyear's purchase of Hayward's patent: "Do you think a man who deliberately patents an invention of another, an honest man?" This was not just false, it was absurd coming from a man who earlier that year had tried unsuccessfully to patent vulcanization based on information directly stolen from Goodyear. In an ironic twist, Day would gain his first U.S. patent that very month, October 1844, by claiming as his own the work of yet another man. In the early 1840s Goodyear had commissioned Amaziah S. Warner, a Springfield machinist and an associate of Horace Cutler, to build a machine that would facilitate the manufacture of shirred goods. Goodyear gave him the specifications and Warner built the machine. When Warner finished the machine, Goodyear couldn't pay. Cutler told Day about the machine. Day bought it for fifty dollars. His subsequent patent for "Machine for Manufacturing Corrugated of Shirred India-Rubber Goods" did not bother to credit Amaziah Warner, let alone Charles Goodyear.

To the end of his life, Hayward felt, with some justification, that history had overlooked his contributions. Posterity, more comfortable with lone heroes, is seldom generous to helpers and also-rans. It could not have been easy for Hayward to see his name relegated to footnotes. But Hayward also knew in his heart that he was not the inventor of vulcanization. After a while his indignation subsided and he began to regret his association with Day. In October 1845 he wrote to Day, urging him to settle the lawsuit with Goodyear. His mollification had a financial angle, to be sure. By now Hayward had a license from Goodyear to manufacture shoes and boots in Colchester, Connecticut. The license, typical of Goodyear contracts, was generous to the holder—Hayward and his partner, Henry Burr, could make three hundred pairs of shoes a day without paying any royalty at all, and paid five cents for each pair thereafter.

Day ignored Hayward's entreaty. He placed advertisements in New York newspapers, questioning Goodyear's patents and claiming that Hayward was the true inventor. In early 1846, Goodyear responded with a lengthy advertisement of his own in the *New York Express*, which was subsequently reprinted in the *Boston Courier*. Goodyear stated his eagerness for the matter to come to court after months of filings and depositions and motions. He promised to produce "whole towns of witnesses" contradicting Day. Finally, Goodyear made clear his frustration at being attacked in New York newspapers while living in New Haven. "As I do not reside in the city, and am, of course, away from the theater of publication, I cannot expect to look after his notorious bulletins *Day* after *Day*."

Shortly before the November trial date, Day abruptly agreed to settle out of court. It could be that Day had a sudden change of heart, realizing he would lose. More likely, Day never intended to go to court at all. Despite the $5,000 fine he paid, the public ruckus in the end accomplished his goals. The results suggest he merely toyed with Goodyear and Suydam.

By the time of the settlement, Suydam, the luckless original licensee, had dropped out of the picture. As part of the settlement, Day received Suydam's license to make shirred goods. In January 1847, a notice appeared in the *New York Fredonian* over the joint names of Goodyear and Day: "The public is hereby informed that all lawsuits between the subscribers have been satisfactorily settled between them; and that by arrangement made, the entire right to manufacture shirred goods under Mr. Goodyear's patent, and to use Mr. Goodyear's compounds in their manufacture, have been vested in Mr. Day." The terms of the contract must have made Day smile. Day would pay Goodyear three cents per yard of shirred fabric he produced. For a relative pittance in settlement fees, plus another $10,000 to buy the license, Day had driven a competitor from the field and walked off with the rights to make himself a rich man.

At the time he offered the settlement, Day's harsh rhetoric toward Goodyear suddenly melted into an odd, obsequious, and short-lived bid for friendship. On November 15, 1846, he invited Goodyear to "make a visit at my house, and enable me to have undisturbed intercourse with you for mutual advantage." In the letter, Day suggested that Goodyear's attorneys wanted to keep the battle going for personal profit. He claimed to have private intelligence from an unnamed "visitor" explaining how Goodyear's lawyers and friends were plotting to wrest the patent from the inventor. Day wrote, "If my senses does not decieve me [*sic*], your old enemies—no, not that—but your old opponents, must save you from your friends. Come to New York at once, and stay here or at Jersey City. . . . The danger that you are in is greater than you are at all aware."

With the settlement sealed, Day immediately published a notice to scare off any manufacturers who might infringe on his license. He offered a $50 reward "to any person who will give such information as will lead to the conviction of any person or concern in the United States, who may be found after this date engaged in the piratical use of his [Day's] patents, or those granted to Charles Goodyear, used in the manufacture of shirred goods." In other words, Horace Day vowed to prosecute anyone who dared to act as he himself had acted.

For all of his late friendliness with Goodyear, Day was not finished stealing from him. No sooner had the settlement been finalized than

Day began slighting Goodyear on the payments of three cents per yard. Nor could Day tolerate remaining within the small but profitable confines of shirred goods, the only goods he was licensed to make. There was simply too much money to be made elsewhere. Soon, he was turning out other vulcanized goods, including "Day's Best Patent Japan Rubber Shoes." The "Japan" was thrown in as a twist on "India," to make his goods seem different from those of legitimate licensees. He also turned out car springs and packing material for heavy machines, in direct violation of Goodyear's patent.

Business was good. Day began casting about for a new factory to expand his manufactures. In 1847, he returned to his hometown of Great Barrington, in western Massachusetts, where he bought an old mill site near a river and promptly made enemies. Even in his hometown, Day relished a fight. Among his first acts was to dam the river to control water flow over his wheel. He either did not contemplate or did not care how this would affect his neighbors, particularly the Berkshire Woolen Company, located just upstream. As water pooled at Day's dam, it gradually backed up to the woolen mill. Backwater sloshed against the woolen mill wheel, rendering it as listless as a becalmed topsail. Furious mill owners demanded that Day dismantle or lower his dam. Day refused. The woolen company sent workers over with sledge hammers to modify the dam. Day rebuilt it. The woolen mill filed suit, and Day refused to back down. The dispute dragged on for years, until a judge at last ordered Day to lower his dam by eleven inches. At the new elevation, water pressure over Day's wheel dribbled too feebly to run his machinery. He packed up and departed, leaving behind a bad reputation and a set of buildings that would molder unused for more than thirty years.

The dispute unfolding in Great Barrington hardly distracted Day from his primary objective: a good, old-fashioned donnybrook with

Goodyear. As his public advertisements for all manner of rubber goods made clear, he had no intention of hiding. But this time, he was taking on not just a disorganized Charles Goodyear and passive David Suydam, but the combined forces of a new, prosperous, and powerful conglomerate of rubber manufacturers known as the Goodyear Shoe Association. The association included the Goodyear Metallic Rubber Shoe Company and the Naugatuck India Rubber Company, both of Naugatuck; L. Candee & Company, of New Haven; Ford & Company of New Brunswick; and the Newark India Rubber Company. The association name implies that Charles Goodyear was the controlling party. In fact, Goodyear had no practical control over the organization—they were simply manufacturers who owned Goodyear licenses. In fact, many licensees would incorporate the Goodyear name into their own. While Charles Goodyear struggled financially and reaped relatively modest rewards from his invention, the proliferation of his name led to the widespread assumption that he was growing rich beyond description. The confusion would reach an apex long after Goodyear's death, with the emergence of the eponymous tire company.

The Goodyear Shoe Association was one of the first trade associations in the United States formed to protect the mutual interests of its members. Price fixing (then considered a sound business practice rather than a crime) was part of the association's mandate. Men's rubber shoes were to cost $1.00 per pair, wholesale; ladies' shoes 75¢. Ladies' long boots cost $2.75, men's top boots $5.00, children's shoes 50¢. But the first order of business was dealing with the growing problem of Horace Day, who by now openly called Goodyear's patent a "vile calumny" and "a fraud and a swindle." Alarmed at his brazenness, members of the association dedicated three cents per pair of rubber shoes manufactured as a communal war chest to defend their licenses. In 1850, Day directly petitioned Congress to repeal the patent. Congress ignored the request, but Day's broadside had its effect—this time, there would be no settlement. The association responded with newspaper notices stating: "All the India rubber shoes now made by Horace H. Day, are an infringement upon Goodyear's patents, and all dealers therein render themselves liable in damages to Mr. Goodyear, the patentee." All this set the stage for what was to become the greatest industrial lawsuit in United States history until that time.

❁ ❁ ❁

In England, Thomas Hancock wasted little time developing practical uses for "his" invention. Hancock at last was ready to descend from his attic laboratory and let others in on his secret. He transferred his experiments from his private laboratory to Manchester, where he could put his heavy machinery and multiple employees to work perfecting the process and finding commercial uses. His first commercial vulcanizing machine involved an iron pan set inside brickwork, large enough to accommodate a sheet of cut rubber. He converted the sheets into buffers for sleeper beds on the Great Western Railway. "The effect produced was a great improvement; but the expense was thought too great to justify its adoption," Hancock recalled. Hancock ran into the same problems Goodyear had confronted in trying to achieve a uniform heat that would vulcanize all parts on a rubber object equally. Like Goodyear, Hancock became frustrated when one portion of a piece would emerge perfectly vulcanized while another would be scorched or barely changed. Goodyear had solved the problem by diverting the heat from the furnace to pipes leading around the oven. Hancock came up with the idea of using steam.

He had experimented with a dry oven, in which goods rotated much as in Goodyear's rotisserie. But Hancock started experimenting with a high-pressure steam boiler, necessarily very small to avoid explosions at high temperatures. The wet heat vulcanized rubber evenly and thoroughly, and seemed to limit the escape of sulfur. He now built a much larger boiler, heavily reinforced. With this new method he could fully vulcanize twelve-inch solid cubes or long sheets of rubber.

Among the early practical uses Hancock found was a method for coating and vulcanizing large canvas sheets, called ship-sheets, for use by the British navy. Some of the sheets were fitted over wooden frames to make small, lightweight, and maneuverable waterproof boats. Ship-sheets also were immediately useful in repairing larger vessels. In the past, ships that sprang even minor leaks often had to be hauled into dry dock so the holds could be drained before workmen could make any repairs. The process was time-consuming and expensive. Now, repairmen could slip a ship-sheet on the outer hull of the ship to cover the leak area temporarily. Water pressure against the hull would hold the sheet firmly in place. Repairmen could then pump the flooded section dry and make their repairs, without the ship's ever leaving the water.

Virtually all of the existing machinery Hancock had used for mixing, molding, rolling, and cutting unvulcanized rubber could be directly transferred to use on vulcanized rubber. This enabled him to get products quickly to market. During the two years after Hancock's patent, Charles Macintosh & Company and its licensees produced an astonishing array of goods. There were washers and packing for steam pipe joints, engine valves, and printers' rollers and blankets. There were billiard cushions, spongy rubber for musical instruments, rubber hose pipe, tubing, and surgical bottles. There were gig springs, carriage springs, railroad springs, and pump buckets. There were gas holders, artificial leather, trouser straps, swimming belts, and horse boots. Rubber was shaped into boot soles, neck pillows, enema tubes, and sitz baths. Now that rubber resisted deterioration, babies could gnaw on teething rings to their hearts' content, maybe even as they hopped in their baby jumpers, suspended from the ceiling by a vulcanized rubber cord. There were fishing trousers, gun covers, inflatable boats, and bellows. In 1846, when Queen Victoria rolled through the streets of London, her royal ride was made smoother by solid rubber tires from the shop of Thomas Hancock.

Among the first Hancock licensees to patent offshoots of vulcanization was a British firm called Perry & Company. In December 1844, Perry & Company introduced a product so humbly indispensable, so unassailably simple, so uniquely utilitarian, that it worked its way into every office and home in the developed or developing world and stayed—the elastic band.

Yet if these were exciting and profitable times for Hancock, he would never be allowed to fully enjoy them. He would never know the peace of mind that comes to a man after working to exhaustion in pursuit of an original idea. Hancock had worked to exhaustion to deprive another man of *his*. Hancock's gains were hard-earned, at least to the extent that he had worked hard to re-create Goodyear's breakthrough, but this did not erase the fact that they were ill-gotten. Even if he could convince himself otherwise, a chorus of detractors, chiefly British, would never let him forget.

The first serious challenge to Hancock's patent came from British importers of American vulcanized rubber shoes. For some reason, rubber footwear had always been more popular in the United States than in England. Hence American manufacturers had developed a special expertise. With vulcanization, rubber overshoes emerged as one of the few products of any kind where American craftsmanship surpassed the English. Hancock, by his own admission, never cared much for the shoe business and devoted little effort to it. His steam-vulcanization method produced shoes that were highly elastic, but rough-surfaced and rather ugly. Consumers preferred the shinier, more polished American finish. Nathaniel Hayward, manufacturing under Goodyear's patent, shipped some shoes from his factory in Colchester, Connecticut, to England in 1847. Hancock sued. To avoid a court fight, a Hayward representative negotiated a deal with Hancock whereby Hayward received exclusive rights to export shoes from the United States to England in exchange for royalties. This deal effectively limited the availability of rubber shoes in England to the relatively paltry number emerging from Hayward's lone Connecticut mill, plus the widely reviled, steam-vulcanized shoes produced in England under Hancock's patent.

English shoe dealers rebelled, openly plotting to overturn Hancock's patent. The strategy was twofold: first, to establish that the American shoes were "infinitely superior" and hence an entirely different product from those covered under Hancock's patent; and second, to discredit Hancock as the inventor of vulcanized rubber. Of course, Horace Day would attempt similar tactics in the United States against Goodyear, with the one significant difference being that the English merchants were telling the truth. Calling themselves the Dealers in American Overshoes, the merchants convened at Guildhall Coffeehouse in London on November 5, 1848. They appointed a committee of six merchants to lead the group. Seeking strength in numbers, they combined all the separate infringement suits against them by Hancock into one massive defense. Any merchant who paid two pounds, two shillings into the common pot could consider himself represented by the group.

At the same time, Hancock's patent was coming under fire on another English front. Though a neophyte when he transported Goodyear's samples to England on that fateful trip in 1842, Stephen Moulton had grown increasingly enamored of rubber, and increasingly convinced of its possibilities. In 1847, Moulton left New York for England, determined to set up a rubber works to challenge Hancock's dominance. He settled in Bradford-on-Avon, a picturesque town in Wiltshire County, western England, a stone's throw from Hancock's hometown of Marlborough. Moulton purchased an old woolen mill on eight acres. The most imposing feature on the property was a grand but faded sixteenth-century mansion known as the Hall. Moulton restored the Hall to its baronial glory (a descendant still lives in the house) and established The Moulton Company, the first English rubber company that challenged the dominance of Charles Macintosh & Company.

As Moulton struggled to learn the trade, he took on as early partners the Rider brothers, Goodyear's old patrons from New York. The Riders had by now recovered from their financial slump of 1841 and had begun manufacturing as a Goodyear licensee in New York. As distant partners in Moulton's English works, they traded advice and ideas and sent over American workmen to help build Moulton's machines. A skilled American mechanic named Henry Frost built a close replica of Chaffee's legendary calender, the Monster; this one was called the Iron Duke.

Dozens of handwritten letters from the late 1840s survive in the Moulton Company archives, most of them from William Rider to Moulton, with some of Moulton's responses. They provide a wonderful source of details on the early rubber industry, such as the required size for a factory with two hundred employees (two hundred feet long, forty feet wide, and four stories tall) as well as the estimated cost of equipment ($19,700). As for Goodyear, the letters offer an irreplaceable glimpse of the paradoxical affection and consternation he engendered in his friends.

Rider and Moulton agonized over whether to include Goodyear in their company. They both liked "Charly" personally and believed that they (and the world) owed him a great deal. But how could they subject their new enterprise to the inevitable cash drain of a man who created gorgeous objects that never seemed to make it to market? Rider wrote from New York in April 1848: "Charly has come here, seems to be determined to associate with us in Europe in the rubber line, and from your account recently, I am inclined to think you will be glad to have such the case. . . . The fact is, Goodyear can do a great many things in a way that no other man can do, besides, he is entitled to a chance with us, for what he has done as a pioneer in the business, but the terms we should give him . . . will be a matter of considerable thought and calculation." Later that month, Rider again raised the issue of including Goodyear "if he will not procrastinate so much." Rider adds, "He has certainly got some things to great perfection recently, as you will see by the samples I send you."

During this time, Goodyear busied himself making more ship sails, plus jewelry cases, globes, and anything else that struck his fancy. "He is now strong on globes, has one three feet in diameter," Rider reported in late May 1848. "When filled with gas they will keep up to the cieling [*sic*] which is very beautiful." A year later: "Goodyear has just arrived in the city and has brought along some very pretty samples in the way of globes, maps and umbrella covers."

Moulton wanted Goodyear in England mainly as a sign of solidarity as he prepared to take on Hancock. Moulton figured that Goodyear's presence on English soil would provide powerful symbolic defiance to Hancock's claim as the inventor of vulcanization. Goodyear promised to visit, and Moulton waited. The letters on this score read like a scene from *Waiting for Godot*. In September 1847, Rider wrote: "Goodyear planning to come to Europe soon." In May 1848: "Nothing new about G., he still says he will go to England now *very soon*." October 1848: "Goodyear coming soon." February 1849: "Goodyear can undoubtedly bring McIntosh [*sic*] to terms if he will but go to London." August 1850: ". . . when he goes to England, if he ever does go." Moulton, equally exasperated, wrote to Rider in July 1848: "Goodyear is coming, of course, so is the year 4673200000000. The latter will arrive first, to be sure."

It's understandable that Moulton wanted Goodyear on hand to bolster his chances against Hancock. Like the shoe dealers, Moulton faced steep odds in his quest to overturn Hancock's patent, because of the nature of English patent laws. English law differed (and still does) from American in one critical way: while Americans recognize the primacy of the original inventor, the creator of the idea, the English, barring evidence of out-and-out illegality, essentially recognize whoever files first for the patent. It is almost reflexive for Americans to howl at the injustice of the English system—what about the rights of the *true inventor?*—so it bears pointing out that neither system is perfect and that each has advantages. The American system, of course, appeals to a sense of fairness since it strives to give credit where credit is due. But the birth record of an idea can be a tortuous thing to establish, especially when multiple parties honestly do arrive at the same idea more or less simultaneously. As inventors and their patent lawyers engage in ever murkier gyrations to prove their primacy, litigation can drag on for years while the public waits to reap the benefits of the invention. English law obviates these difficulties (or seeks to, anyway), by putting the onus on the inventor to keep his secrets and get to the patent office with due speed. The system may seem ruthless, but it is cleaner than the American system and helps ensure that new ideas reach the public quickly. Unfortunately, it also clears a wide path for the underhanded.

And so the shoe dealers and Moulton, in their separate cases, essentially had to prove the unprovable: not that Hancock lifted the idea, but that Charles Goodyear had, in fact, beat Hancock to the patent office. And everybody knew that wasn't so. The only hope was to portray Hancock as utterly ignorant of his goal when he filed for his provisional patent in November 1843. Then Goodyear, who applied with full details eight weeks later, might emerge as the first to *credibly* apply. But since provisional patents were intended precisely to protect ideas in their formative stages, attacking Hancock for not having the answers was a long shot at best. About the best Hancock's enemies could realistically hope for was a moral victory. Perhaps that would suffice.

The shoe dealers' case came to court first, in 1851. Moulton appeared as a witness on their behalf, recounting in detail how the samples wound up in Hancock's hands, how he had met with the Birleys and made clear to them that the honor of Charles Macintosh & Company rested on their respecting Goodyear's confidence. It must have been agonizing for Hancock to hear the story told in so public a forum. More agonizing still, Hancock soon found himself on the witness stand. Under oath, and under harsh questioning, he acknowledged that he had in fact seen Goodyear's rubber samples before launching his own quest. The world's most renowned rubber man, the father of the industry, had to admit that he had acquired secondhand the one idea that mattered most.

As expected, the shoe dealers lost. Hancock's patent was reaffirmed as lawful, given the narrow requirements. He still had the patent, but he also had the unbridled contempt of his enemies. As George Ross, a leader of the shoe dealers' group, wrote to Moulton: "That [Hancock] obtained whatever knowledge he possesses of Vulcanizing India Rubber from *American* Vulcanized goods . . . has been uncontestably [*sic*] proved." Hancock probably cared little for the opinions of shoe dealers. As for the Moulton case, that battle would twist and turn for another five years before finally seeing a courtroom resolution in 1855 before the Court of Queen's Bench. This would ultimately bring assessments of Hancock's character that he would find harder to shrug off.

Meanwhile, Hancock's legal victory over the shoe dealers may have prevented a wave of Goodyear-inspired, American-made rubber shoes from flooding the British markets. But his acknowledgment of seeing the samples from America provided a long-awaited moral victory for Goodyear. When Goodyear did at last come to England, it was to launch another sort of invasion to stake his claim to the throne of preeminence in rubber. And Hancock could do nothing to stop it.

❁ ❁ ❁

Queen Victoria officially opened the Great Exhibition in London's Hyde Park on May 1, 1851. The full title was the Great Exhibition of the Works of Industry of All Nations. It was a bold, dramatic embrace of science and industry, staged by the greatest industrial power. Never had the world seen such a display. The exhibition drew seventeen thousand exhibitors, all of them housed within the iron-and-glass walls of the Crystal Palace, designed by the architect Joseph Paxton, who was selected from more than two hundred applicants. There were looms from Belgium, armaments from France, vases from Russia. There were clocks from England, lace from Spain, and agricultural equipment from the United States. It was a time for displaying all that was new, exciting, and beautiful in the modern world.

Here, at last, Goodyear found a forum commensurate with his own romantic conception of rubber. With the exhibition's premium on the aesthetic and dazzling over the economical and money-making, Goodyear was in his element. Here, he would not be measured against his ability to turn a favorable balance sheet, but by his ability to create beauty through rubber. The gorgeous objects he had slaved over, to the consternation of his investors and partners, finally found their venue. The stall housing the Goodyear Vulcanite Court was divided into three rooms. The partitions were made of hard rubber, shiny and dark with elaborate scrollwork. Goodyear had spent months creating and gathering his favorite pieces for the exhibit. Now, hobbling around on crutches because of painful flare-ups of gout, he tirelessly directed every part of the installation. It was a magical display. Just inside the entrance sat an ornate desk made entirely of hard rubber, decorated with gilt around the edges. There were portraits painted on slabs of rubber, waterproof maps, soft creamy curtains of rubber-covered cloth, and large armoires made of hard rubber. He brought along some of his beloved ship sails, too. The smaller objects were no less notable. There were bracelets, syringes, medical instruments, walking sticks, rubber heads for dolls, and exquisite boxes inlaid with stones. Hydrogen-filled rubber balloons, painted with the nations of the world, danced about the ceiling. His work bore the stamp of an electrifying, if slightly mad, genius at the height of his creativity. Appreciative crowds thronged the Goodyear exhibit.

Thomas Hancock also created a display, located in a different part of the Crystal Palace. The responsible, businesslike Hancock filled his stall with Charles Macintosh & Company products: coats, tarpaulins, solid carriage tires, rubber boats, and the like. The most fanciful object in Hancock's display was a round slab, about the size of a dinner plate, featuring the molded likeness of his own face. It is hard to imagine that the foes didn't take time to peek at each other's exhibits. Goodyear's heart must have swelled with the comparison. Hancock's must have skipped a beat.

"A visit to the Crystal Palace, and an examination of the contents of the stall of Mr. Goodyear cannot fail to create astonishment," the London *Art Journal* noted, in an article that did not mention Hancock's exhibit. "For while it will be seen how numerous are the objects already wrought out of India-rubber, these will only convey an idea of the still larger number of uses to which it is destined to be applied."

The Goodyear Court struck the *Art Journal*'s correspondent as "costly in character," and this was an understatement. Goodyear had spent thirty thousand dollars on the exhibit, much of the money borrowed from William DeForest. Any windfalls from his licenses were blotted out, replaced by a new generation of debt. Goodyear had convinced himself and others that his extravagance had sound financial underpinnings: the exhibit would so astound the world that governments from many nations would clamor for his products and expertise. But he had no formal plan to reap this expected harvest. Most of the goodwill he generated drifted off like one of his gaudy balloons let loose in a summer sky.

But who could think of such things at a time like this! Goodyear had pulled off an aesthetic triumph, right in Thomas Hancock's backyard. Both men would come away with high medals for their exhibits. But where Hancock had made a solid, conventional statement of industrial prowess, Goodyear had shown the world (and Thomas Hancock) that no man could match his artistry, imagination, passion, and genius when it came to rubber. What was mere money compared with that?

11

WEBSTER, CHOATE, AND A SLEEPWALKING MURDERER

Goodyear had only a few short months to bask in the glory of London before his impatient licensees called him back to the United States in late 1851. They needed him to help prepare for the impending battle with Horace Day. The Day case was shaping up as a landmark trial, one that the licensees hoped would not simply settle matters with that irksome character, but would also help solidify the general power of patent laws to protect inventors and licensees. Rampant infringement had been a nettlesome problem from the time the U.S. Patent Office was established in 1790. Eli Whitney patented his cotton gin in 1793, only to find himself chasing copycats in court case after court case throughout the South. Most of the $90,000 he earned from his invention went toward legal fees. In those days, infringement cases played out in local courts, so Whitney was forced to travel to small towns in cotton-growing states to sue bogus manufacturers before homegrown juries. Federal courts had by 1851 taken over infringement cases. But an inventor still faced steep and often insurmountable odds in protecting himself against multiple violators, who usually financed their defenses with ill-gotten proceeds—the fruits of the inventor's own creativity. "This is, to the inventor, a grievous hardship and wrong, and has no parallel in any other species of property," Goodyear noted bitterly in his memoirs.

To prosecute their case against Day, the Goodyear Shoe Association set its sights on the greatest orator and most celebrated lawyer of the age. Today, Daniel Webster may be best known as the eponymous character in *The Devil and Daniel Webster*, Stephen Vincent Benet's fanciful play from the 1930s. Webster the actual man rattles around in the attic of our national memory as a famous but oddly hard-to-place New England statesman, who gave important speeches on matters that, like the Whig Party he once represented, have long since receded into the historical haze. Many people assume he had something to do with dictionaries. He did not. He and Noah Webster, the Connecticut-born dictionary man, were unrelated.

In his own time Webster was a celebrity on the order of a modern movie or rock star, but with intellectual underpinnings. In an age of oratory, of four-hour speeches and three-day trial summations, Webster had few rivals and no superiors. He was a patriot, a defender of the Constitution, and an advocate of a strong federal government. When in 1820 he commemorated the bicentennial of the landing of Pilgrims at Plymouth Rock, his address drove women to tears and cynics to thump their breasts in patriotic fervor. His speeches began slowly and built to crescendos that audiences compared with thunder. Printers snapped up copies and sold them like potboilers. He was physically imposing, with dark features, wild tangles of black hair, and deep, full eyebrows that meshed when he was angry into a single menacing ridge. His fellow undergraduates at Dartmouth College nicknamed him Black Dan. The name stuck with him for life. His constituents had another name for him, one they invoked without a trace of irony or reserve: the Godlike Daniel Webster.

The son of a New Hampshire farmer, Webster rose from rural obscurity to claim a pivotal role in the great political and judicial debates of his age. He had been a congressman from New Hampshire and Massachusetts and a senator from Massachusetts, and had argued 223 cases before the Supreme Court. A passionate advocate of the Union and of a strong federal government, Webster waged celebrated battles against states' rights advocates. In an 1830 speech from the Senate floor, he fixed a stare on Senator Robert Y. Hayne of South Carolina and thundered, famously, "Liberty *and* Union, now and forever, one and inseparable!"

Only the greatest political prize, the presidency, had eluded Webster. After two unsuccessful runs as the Whig candidate, he found himself in 1851 approaching his seventieth birthday, serving out his political career in the anticlimactic role of secretary of state to his intellectual inferior, Millard Fillmore. He had lost a step to age and various ailments, among them cirrhosis of the liver from decades of heavy drinking. His voice lacked some of its former power. And he had lost some support among Massachusetts abolitionists, despite his lifelong opposition to slavery, for declaring that preserving the Union was more important than immediately freeing the slaves. Among a broader public, though, Webster had long since achieved a sort of mythical status, something beyond that of a mortal man, beyond the bounds of any single speech or issue. He was widely regarded as a national treasure by a populace careening toward civil war and longing for the clear, hopeful patriotism of Revolutionary days. They saw in Webster a living bridge to that idealistic time. Born in 1782, he was too young to have joined the Founding Fathers in Philadelphia. Yet his early career intersected with the ends of theirs. He had known the gods. They liked and respected him. A young Webster traded political ideas with James Madison over dinner. He spent five days as a houseguest of Thomas Jefferson at Monticello, Jefferson's hilltop estate in Virginia. He attended the funeral of John Adams, and delivered a two-hour eulogy for him in Boston a month later.

That the members of the Goodyear Shoe Association would even consider approaching so august a figure to represent them in court seems an astonishing act of hubris. Indeed, Webster's first reaction was to turn them down cold. This was a major business case, to be sure. But Webster hardly needed the publicity to be had from representing a bunch of shoe manufacturers. What changed Webster's mind was a reality as old as civilization, and something that Charles Goodyear, of all people, could understand: He needed the money. The godlike Daniel, orator of the ages, houseguest of Jefferson, and scion of Marshfield, Massachusetts, was broke.

For men with few tastes and habits in common, Goodyear and Webster did share one trait, an abject disregard for money. They believed that the power of their ideas would shield them from the need to balance a checkbook, and were always surprised when this proved not to be the case. Goodyear's weakness was rubber; Webster's, luxury. Webster, who ought to have been a wealthy man based on a lifetime of generous fees and court settlements, frittered away his earnings on bad investments and high living. He poured money into his grand country estate in Marshfield, and treated his incessant stream of houseguests to the finest vintages and delicacies. Webster's string of creditors rivaled Goodyear's.

The Shoe Association must have known about Webster's plight or they never would have summoned the courage to approach him. The resurgence of the rubber trade, thanks to Goodyear, had swollen the association's defense fund. They offered ten thousand dollars as an initial payment, plus five thousand more should he win. As an additional incentive, George Griswold, a New York rubber manufacturer, offered to chip in another thousand for a win. The total fee—ten thousand dollars, guaranteed, sixteen thousand total for a victory—was the largest ever dangled before an American lawyer for a single case. Webster simply could not refuse.

Not to be outdone, Horace Day selected as his own attorney Rufus Choate, a figure now largely forgotten, but in his day nearly as famous as Webster. In fact, the two had much in common. Choate reminded people of a younger Webster. Like Webster, Choate, seventeen years his junior, was a New Englander. As youths, both had retreated to the comfort of books—Webster due to chronic childhood illness, Choate because of an almost complete lack of interest in sports and physical activities. Their reading positioned both for precociously brilliant careers at Dartmouth: Webster graduated Phi Beta Kappa at nineteen, Choate graduated at the head of his class when he was twenty. They shared brooding physical features that added to their powers of persuasion. Choate idolized Webster; Webster respected and admired Choate. Their paths had crossed frequently. They met in court, sometimes arguing as partners, in other cases as adversaries. Both served in Congress from Massachusetts, and when Webster left the Senate in 1841 to serve as secretary of state under William Henry Harrison, it was Choate who assumed Webster's seat.

Webster's name inspired more awe among the citizenry than did Choate's, but the shrewd Horace Day had to like his chances going into battle. Webster was old and tired, clearly past his prime. Choate was a vigorous fifty-two, still at the height of his skills. Even without the age difference, many observers would have given Choate the edge. If Webster was the superior statesman, speechmaker, and Constitutional scholar, Choate was a more celebrated master of the courtroom. Admirers called him Wizard of the Law, and the Great American Advocate. He had a rare ability to connect with juries, to communicate in language that was plain but not condescending, and he frequently won cases in which nobody gave him a chance. Choate's reputation for courtroom genius was sealed when he established sleepwalking as a defense in one of the most sensational murder trials of the nineteenth century.

In late October 1845, the residents of Cedar Lane in Boston had been enjoying a glorious stretch of Indian summer, sleeping with their windows thrown open. They were woken early one morning by a scream from a two-story boardinghouse "of ill repute," as one newspaper would describe it. A few moments later, residents of the house heard a man rush down the stairs, fall, get up, and run out the door.

The owners burst through the door of a second-floor apartment to find a smoldering fire and the body of Maria A. Bickford, a beautiful, twenty-three-year-old tenant. Miss Bickford lay on her back, a bloody sheet covering her body but leaving exposed her head, breasts, and feet (as the newspapers breathlessly reported). Her head was turned awkwardly to the right and thrown back, exposing a horrible gash. The coroner reported that the murder weapon—a shaving razor, found bloody and opened next to the corpse—had cut cleanly through the victim's jugular vein and windpipe and touched bone, nearly severing her head. The walls were splattered with blood from the violence of the cut. A basin was filled with bloody water, where the killer had hastily washed his hands. He had gathered a pile of clothes on the bed and lit them. Matches lay strewn about the room. Maria wore a ring inscribed "To MAB from AJT."

Albert J. Tirrell, of Weymouth, Massachusetts (where he had a wife and two children), was the immediate and overwhelming suspect. He was arrested several weeks later in New Orleans, after a flight through several southern states. Among the objects found in Maria's room were a letter from Tirrell to Maria, as well as a cane, underwear, socks, and keys belonging to the man. To top off the prosecution's case, Tirrell and Maria had been heard arguing violently the evening of the murder. By the time prosecutor S. D. Parker opened his case against Tirrell in late March, the public and the newspapers practically had a noose around the defendant's neck. Outside the courtroom, scalpers hawked passes. The courtroom was jammed each day. The compelling prosecution witnesses included the owner of the Boston stable where Tirrell had arrived around dawn on the day of the murder, looking dazed and demanding a ride from Boston to Weymouth.

The only point in Tirrell's favor was that he had recently inherited some money and therefore the means to hire as his attorney Rufus Choate. S. D. Parker watched in dismay as Choate punched holes in his airtight case. First, Choate suggested that someone else might have sneaked into the room and killed Maria, after Tirrell left. Then, he suggested that Maria took her own life. In order for this scenario to square with the undisputed forensic evidence from the scene, Maria would have had to set fire to her room, nearly sever her own head with a razor, and then, having sliced through her jugular vein, stand at a basin and wash her hands before lying down to die. This still would not explain the man heard stumbling down the stairs after the scream. As Choate spun the suicide scenario, the prosecutor looked to the jury box, expecting to see members straining to contain their guffaws; instead, he saw expressions of respectful curiosity. They didn't call Rufus Choate Wizard of the Law for nothing.

But Choate saved his best for last. Perhaps Tirrell did, in fact, do away with Maria. But how could a man consciously do such a thing to a woman he loved? Was it possible he was himself a victim of *somnambulism*? Choate introduced family members who insisted Tirrell was a chronic sleepwalker, as well as physicians who affirmed that somnambulism was a bona fide medical condition. He read from *Harvey's Meditations* a story of two hunters sleeping in the same room. One of the hunters, dreaming about killing a stag, grabs a knife and plunges it into the heart of his friend, waking only after the friend lies dead. With a rhetorical flourish, Choate fixed the jury with an incredulous stare and asked if the hunter could be deemed a murderer and held morally responsible for a crime. Was the poor soul not, in fact, simply *another victim*? Of course, Tirrell's sleep would have had to be extraordinarily deep, since it allowed him to light fires, wash his hands, stumble down the stairs into the night, and hire a carriage to take him back to Weymouth at dawn.

A nervous judge instructed the jury to "take great caution in a defense of this kind." And yet the jury required just two hours and ten minutes to return with a verdict: not guilty. One prominent critic, Wendell Phillips, angrily wrote that Choate had "made it safe to murder." The jury, chagrined at the criticism, later denied that they had considered the sleepwalking defense. But Rufus Choate's reputation as a legal miracle man was assured that day. And whatever an outraged public thought of Choate's defense, the wily Horace Day considered him just the man to help him prevail over the massed forces of the Goodyear Shoe Association and Black Daniel Webster.

Hundreds of spectators made their way to the U.S. Circuit Court in Trenton, New Jersey, for the opening of the trial on March 23, 1852. The state would not see another of such magnitude or fanfare for another eighty years, when Bruno Hauptmann faced trial for the Lindbergh kidnapping. Nobody wanted to miss the action. Journalists poured in from as far away as Boston. Unable to find seats, many of them were forced to file reports based on interviews with spectators as they left the building. Outside, vendors sold refreshments to the milling crowds who gathered simply to be near history in the making. The state legislature shut down for the week so that legislators could watch the hearings and jockey for proximity to Choate and, especially, to Webster.

When Webster arrived from Washington, thirty-seven legislators and members of the New Jersey Bar offered to hold a banquet in his honor following the trial. Webster declined, citing the demands of the case and his need to return to Washington immediately after the case concluded. Instead, he agreed to visit the floor of the New Jersey legislature for a special ceremony in his honor. Before a joint session, Webster gave a moving, patriotic, diplomatic speech in which he spoke of his "hallowed regard" for New Jersey's contributions to the American Revolution. Senator Richard Stockton, a native of nearby Princeton and commodore of the United States Navy, introduced Webster, saying, "If there is a patriotic heart that warms the bosom of any man, that heart is in the body of Daniel Webster."

Crowds were so thick that after the first day of hearings in the weeklong trial, the site was changed to the more spacious Mercer County Courthouse, with seating for seven hundred. Even in the new venue, spectators vied for seats. Webster's frailty and sickness hardly detracted from the impression he made. The *Trenton State Gazette*'s description was typical: "Intellectual greatness will always assert and maintain its own proper rank. Notwithstanding all the diversities of political sentiment, Mr. Webster is considered by the people of the United States the master genius, not only of the country, but of the age."

Not all of the excitement centered on the celebrated personages. The scene gave dramatic testimony to the rise in prominence and importance of rubber in the eight years since Goodyear in the United States and Hancock in England had patented vulcanization. Now companies fought for the right to produce rubber goods. From the start, journalists and judges alike took to calling the Goodyear-Day lawsuit the Great India Rubber Case.

Both sides had taken steps to inflame passions and shift public opinion in their favor. William Judson, one of the Goodyear attorneys, hired a writer named D. F. Bacon to pen a series of attacks on Horace Day, which were published in a variety of newspapers prior to the court hearings. Day had launched his own broadsides against Goodyear and the licensees. This time the sides were going to war, and they knew it.

In preparation for the trial, both sides spent months and thousands of dollars taking depositions from hundreds of witnesses, who did their best to recall minute events that were more than a decade old.

Goodyear's attorneys set a strategy of emphasizing Goodyear's struggles and suffering, to heighten sympathy toward him and, they hoped, anger at Day. The attorneys interviewed more than a dozen residents of East Woburn.

Elizabeth Emerson, the Goodyears' erstwhile landlord, remembered the family's suffering through the long winters, digging half-grown potatoes for food, and how the kindness of neighbors kept the family from starvation. She also recalled, with a mixture of condemnation and respect, witnessing Goodyear's single-minded devotion to his mission when the family boarded briefly in her home. "He used to bring home pieces of India rubber every evening," she said. "And after we got through our work in the kitchen, he used to take possession of our stove in experimenting, in heating rubber. I have seen him put pieces of rubber in the oven of the stove, and when he took them out, he used to nail them up in the house; he was constantly bringing home different pieces of India rubber and trying them in various ways, putting them into the stove, and on the stove, and in the fire shovel, into the fire; he used to prepare some kinds of composition and heat them with rubber on the stove; he used, after heating them, to take strips of India rubber and try to spin them into cords; and he was constantly experimenting in different ways. . . . He began this as soon as he came and continued it as long as he stayed; he was always engaged so when he was at home, and often till twelve o'clock at night."

Cobblers William Trotman and Herbert Beers recalled making rubber shoes that were sacrificed to the fires of Goodyear's obsession. James McCracken painted a portrait of a ragged, pathetic Goodyear tramping into his shop bearing samples and begging to hold them in front of the fire. As Goodyear sat silently watching the court proceedings, his gaunt, haggard appearance gave mute corroboration to the testimony: here was a man who had suffered for his passion.

The defense, headed by Choate and Francis B. Cutting, made no attempt to deny that Day had infringed on Goodyear's patent. Everybody knew that he had, willfully and often. Instead, the strategy was to attack the validity of the patent itself by attacking Goodyear's credibility as the inventor of vulcanized rubber.

Although Nathaniel Hayward had already disavowed any claims that he had invented vulcanization, Day's forces charged ahead with the proposition that Goodyear had stolen the idea from Hayward. But they didn't stop with Hayward. For months leading up to the trial, Day and his lawyers scoured the East Coast, poking into every nook and cranny for would-be inventors willing to claim credit for having come up with vulcanization before Charles Goodyear. The vulcanization process may have taken Goodyear eight years and countless setbacks to discover and perfect, but according to Day's attorneys, it was already old news. Day's roster of inventors had done remarkably little to publicize their discovery until questioned by lawyers in the Great India Rubber Case.

Foremost among these claimants was an obscure, unemployed machinist named Richard Collins, elevated by Day's attorneys to the status of a great and unjustly neglected inventor. Collins claimed that while living in a single room at Budlong's boardinghouse on Eutaw Street in Baltimore in 1833, he purchased several pairs of rubber shoes and decided to see if he could turn them into packing material for steam engines. In the privacy of his room, he mixed sulfur, white lead, magnesia, lampblack, and turpentine. Recall that 1833 was about the time Charles Goodyear had his first serious encounter with rubber, at the New York office of the Roxbury India Rubber Company, when he stopped in to buy a life preserver. If we are to believe Richard Collins, the next ten years of Goodyear's life were wasted. Collins, on his first try, happened not just on the correct ingredients, but on the proper heating method, to produce vulcanized rubber—the process that for years had eluded not just Charles Goodyear but also Nathaniel Hayward, Thomas Hancock, and any number of other diligent rubber men. He had not so much as a single false step. Collins said he built a cylinder of sheet metal, which he suspended in a cast-iron stove that he said had been in the room when he rented it. He placed a piece of duck cloth coated with his rubber mixture into the cylinder and left it to bake over a moderate fire overnight: "In the morning I took the cylinder out of the stove-oven, and took out one head, and finding my rubber cloth had done well, and appeared well, and was quite hard." He tried the experiment three times and got the same results each time, he said.

Collins moved to Lowell in 1834, where he claimed to have repeated his experiments, with astonishing success. He used the rubberized cloth to make two pairs of pantaloons, which he sold to "two gentlemen" whose names he could not remember. He told no one of his experiments except for his brother, John. In 1835 he went to work for the Boston Steam Factory as a machinist. While the company manufactured rubber goods using the conventional (and unsuccessful) formulae of the time, Collins said he and his brother, John, continued on their own to work with sulfur and white lead. Collins said he dissolved rubber with turpentine in tin cans. When the rubber was softened, the Collins brothers would cut it into thin slices and sprinkle it with sulfur and white lead. The rubber would then be mixed again, and spread onto cotton drilling cloth, which they fashioned into overcoats, Collins said. They continued this until the factory burned down in 1836. Later, he went to work for a Salem manufacturer where, he claimed, he sold the process to three men whose full names he could not remember. Under cross-examination, Collins testified that the rubber, sulfur, white lead, and heat method he used in his Baltimore room was the first thing he ever tried. Nor did he encounter any of the agonizing difficulties Goodyear had endured when goods emerged from the oven imperfectly vulcanized, either burned black throughout or crispy in spots and raw in others. To hear Collins tell it, his items came out "tolerably smooth" and a dark, even color.

About the best that can be said of Richard Collins is that he was an inexperienced and therefore clumsy liar. Had he possessed the unctuous skills of Horace Day, he would doubtless have injected a few false starts into the history of his rubber odyssey. He would have preceded his moment of Eureka! with at least a month or two of dead ends. The idea that a machinist could buy a few pairs of rubber shoes, take them back to his rented bedroom, and unlock a great industrial secret, stretched credulity well beyond the breaking point. And then there was the matter of his continued experiments in Boston, Lowell, Salem, and Newton. Fellow employees, called by Goodyear's lawyers, testified that they never saw Collins mix rubber with sulfur or lead. William Lovell, foreman of the rubber grinding and mixing department for the Boston Steam Factory (where Collins claimed to have made overcoats from his rubber-sulfur-lead mixture), said, "We used spirits of turpentine and lampblack; pure spirits of turpentine and lampblack. That is all the ingredients while I was there." According to Lovell, Collins was hired to install new rollers, and "was never employed in mixing the ingredients, to my knowledge." Isaac Pierce, who had the keys to the private mixing room, reiterated Lovell's testimony. Henry Blake, who later worked with Collins, coating cloth with rubber out of rented rooms in a brick factory in Salem, said he and Collins did all the mixing together and never used anything but turpentine, lampblack, and magnesia. Of course, Collins claimed that he and his brother only used sulfur and lead when others weren't around. But it is hard to imagine that the brothers Collins were able to mix and cook anything involving sulfur, in the confined space of a mixing room, without coworkers catching a whiff. Sulfur leaves an unsubtle calling card, a nauseating aroma of rotten eggs. Asked if he would have been able to detect the aroma had Collins indeed experimented with sulfur, Blake responded dryly, "I think I should."

Day's attorneys produced several other independent inventors of vulcanization, each about as convincing as Collins. Harrison Perien, of Ellington, Connecticut, claimed to have manufactured shirred goods using rubber, sulfur, and white lead starting in 1843. Elisha Pratt, a Salem hatmaker, swore that he made hats that he coated with turpentine, asphalt, litharge, lead, and sulfur, and then heated. He claimed to have done this in 1834 or 1835. "I found a great improvement; it made it very elastic and springy," Pratt testified.

But the most astounding claim made by the defense was that the first discoverer of vulcanization was, in fact, none other than Horace H. Day. Day claimed that as early as 1827 he had mixed rubber, lead, and sulfur, spread the mixture on shoes, and baked it. Ford Skinner, a shoemaker employed by Day's uncle, Samuel, swore that he'd seen Day do just that. Day would have been fourteen years old when he conducted these experiments. Certainly no one but Day would have dared to make such a claim, and no defense operating without the exalted confidence of Rufus Choate would have dared to suggest it in court.

Webster's presence in the courtroom added weight to Goodyear's side, but he spent most of the early portion of the trial making notes for his summation. Other lawyers for the Goodyear team called a string of witnesses to dispute these so-called inventors, and methodically laid out the pattern of Day's infringements. They traced his initial infringement on the shirred-goods patent Goodyear had extended to David Suydam, the subsequent settlement between Day and Goodyear, and Day's duplicity in disrespecting the deal. They demonstrated that Day had grown rich off the shirred-goods patent even as he cheated Goodyear. The attorneys pointed out the illogic of Day's claiming that Goodyear's patent was invalid, six years after signing a license agreement based on the very patent.

One of the Goodyear lawyers' most compelling arguments recalled Day's initial settlement with Goodyear years earlier, when Day had paid a fine and agreed to restrict himself to making shirred goods under license from Goodyear. After the settlement, Day placed prominent notices in several newspapers, warning any and all would-be manufacturers from infringing on the license, and threatening legal action against violators. The lawyers asked a logical question: If Day considered Goodyear's patent valid enough to protect it just a few years earlier, how could he now claim that the patent had never held any viability? Any reasonable assessment gave the verdict to Goodyear, simply based on the evidence. But all would depend on the closings of Choate and Webster, who, until now, had left most of the heavy lifting to their assistants. A mountain of details had been sifted through in the trial's opening days. Now it all came down to the clash of the titans.

12

BLACK DAN SPEAKS

Before the trial began, Rufus Choate had argued strenuously for a jury, in hopes of playing to his well-established touch with the common folk. Certainly, a man who had presented Albert Tirrell's somnambulism defense would have relished open, attentive faces in the jury box when it came time to present the likes of Richard Collins as the first man to vulcanize rubber. But the two experienced judges assigned to the case, Robert C. Grier and Philemon Dickerson, rejected Choate's request, deciding that they alone would hear the case.

The fourth day of the trial, a Friday, belonged to Choate. He spoke for five hours, from nine A.M. until two P.M. According to eyewitness accounts, the audience sat rapt throughout. The only interruption came when Webster rose from his seat with a well-staged flourish just before noon and left the courtroom surrounded by a train of admirers to keep his appointment with the New Jersey legislature. Unfortunately, nothing remains of Choate's oration. It's unclear whether Choate spoke from prepared text or notes, or whether anyone in the courtroom recorded his speech. No copy remains among the voluminous papers he left behind after his death in 1859. Newspaper accounts were little help, assuring readers that Choate's speech was magnificent, but not bothering to quote from or even summarize the contents. Undoubtedly Choate dug away at Goodyear's claim to sole inventor status, and also did his best to paint Day as a victim of Goodyear. In their 1846 agreement, which gave Day the shirred-goods license, there was a rather nebulous provision stating that Goodyear must "protect" Day from outside infringement. According to the defense, Goodyear failed to offer such protection, thereby negating the contract. Choate's speech impressed observers as vintage Choate, vigorously argued, articulate, and passionate.

The *Boston Courier* reported in the next day's edition: "The great argument of Hon. Rufus Choate, in behalf of Mr. Day, has just been closed. Never was such an argument made in New Jersey. Of all the great efforts his speech this day will stand first and foremost. The various courts sitting in Trenton have drawn hither the leading members of the bar from all parts of the state. More than one prominent member of the New Jersey Bar said in our hearing that the argument to-day was the most able they ever heard. The large courtroom was crowded, and the friends of Mr. Day appear in fine cheer." Unfortunately, the smitten reporter didn't see fit to pass along a single word to his readers. Quite possibly he was one of those unfortunates locked out of the overflowing courtroom.

When Choate finished, Francis Cutting, his associate, launched an extended recitation that lasted the balance of Friday, plus all of Saturday and Monday, presenting the cases of Collins, Elisha Pratt, and the others who, according to the defense, had preceded Goodyear in the vulcanization of rubber. Cutting also pushed hard at the relationship between Goodyear and Nathaniel Hayward, suggesting that Goodyear had simply stolen Hayward's invention when he bought Hayward's patent for sun-drying rubber and sulfur together.

Now there was but one voice to be heard from. Daniel Webster sat at the plaintiff's table next to Charles Goodyear on Tuesday morning as the overflow crowd noisily rustled into their seats. Spectators had come from as far away as New York and Massachusetts to hear the master.

Webster and Goodyear must have presented a touching scene sitting side by side, two grizzled old warriors. Webster was two months past his seventieth birthday and looked every year of it: tall, his unkempt tangles of once jet-black hair were thinner and graying and framed a sallow, craggy face. Goodyear was eighteen years Webster's junior but was his contemporary in the ravages of disease and time. He was slender and stooped. His hair was shorter and straighter than Webster's, but equally unkempt, sprouting like a patch of crabgrass from the top of his head. Goodyear's face, like Webster's, looked almost worn out, the skin stretched tight over his prominent forehead and cheekbones. Goodyear, approaching his fifty-second birthday, was showing the cumulative effects of worry, poor diet, illness, and constant exposure to any number of chemicals. If the presiding jurists, Grier and Dickerson, had any room in their legalistic hearts for sentimentality, they could not have helped but be moved by the site of Goodyear and Webster together. Their opponents across the room, the youthful Day, still shy of forty, and the confident, robust Choate, presented a striking contrast of vigor, one that did little to establish Day's case as the victim.

As was his custom, Webster waited until the crowd had grown completely and respectfully quiet before he rose. This could not have taken long, despite the number of people packed into the cavernous room. People feared the transfixing power of Daniel Webster as much as they admired it. It was part of his legend. Men who served as clerks to Webster in their youths would write memoirs in their old age, still recalling with fresh terror having committed some trifling offense that Webster deemed not properly respectful, and being fixed with one of Webster's brow-knit glares. Even in a crowd of seven hundred, nobody wanted to be the one to make some noise that would cause Webster to wheel around and stare.

James T. Brady had laid out most of the details during his own arguments the previous week, parsing the depositions and attacking

Day point by point. Webster was no expert on rubber, patents, or industrial law. Everybody, including Webster, knew exactly why he was there—for his star power and the theatrical impact of his speech. But Webster's speech showed that he had listened carefully when briefed by Brady and Goodyear. And he had done considerable reading. This was to be no mere star turn.

Webster rose slowly. In keeping with his style, he started in a deliberately dry, low-key way, apologizing for taking up the court's valuable time for it no doubt had better things to do than listen to an old man carry on: "If I should detain the court by the part which I have to perform in this discussion for any length of time, I hope the court will believe that what I have to say is long only because I have not had time to make it short." Appreciative laughter rippled through the audience. As if seven hundred people inside the courtroom and hundreds more outside hadn't come all this way just to hear him speak.

The Constitution was Webster's lodestar; he assessed every issue or idea in relation to that document. So naturally enough he laid the foundation for his Goodyear address with a layer of constitutional theory on individual property rights. He made an eloquent and persuasive case for the still somewhat novel concept, in a nation of manual laborers, that an idea was a form of property as surely as a house or a piece of land. An idea may be stolen as surely as a purse, he suggested, for "what a man earns through study and care is as much his own as what he obtains by his hands."

Webster next took pains to establish for all present the right of Congress to award patents. It was a neat rhetorical trick, for it suggested by implication that Horace Day, with his spurious attacks, had questioned not just Goodyear but the highest lawmaking body in the land. Day was not just a liar and a thief, Webster seemed to imply, but perhaps something short of a true American as well.

Webster continued with a concise and well-organized history of the rubber trade, starting with the first European contact with the rubber tree to the boom and collapse of American factories during the 1830s: "All these factories, all the earnings of individuals connected with the business, came to an end and nothing was done until the business was revived by Goodyear's great invention of vulcanizing. . . . All these prior inventions became non-entities."

Webster recounted some of the problems with unvulcanized rubber, relating his own story of the friend who sent him a rubber cloak and hat. When Webster left the coat and hat out on his porch on a winter evening, the next morning they were stiff enough to stand on their own: "Many persons passing by supposed they saw standing by the porch, the farmer of Marshfield." The crowd laughed.

With his audience primed, Webster segued into a lengthy discussion of Goodyear's sufferings, which the inventor's own sickly appearance in the courtroom affirmed nicely. One observer who attended the trial recalled later, "Charles Goodyear sat at the end of the court table where, to the right, Webster stood. Mr. Goodyear's face showed all the suffering he had undergone." Webster said, "It would be but painful to speak of his extreme want—the destitution of his family, half clad, he picking up with his own hands, little billets of wood from the wayside, to warm the household—suffering reproach—not harsh reproach, for no one could bestow that upon him—receiving ridicule and indignation from his friends." Webster spoke of the long-suffering Clarissa, saying, "In all his trials, she was willing to participate in his sufferings, and endure everything, and hope everything; she was willing to be poor; she was willing to go to prison, if it was necessary, when he went to prison; she was willing to share with him everything, and that was his only solace." There was, he told the audience, "no creature who for the object of her love is so indomitable, so persevering, so ready to suffer and to die."

Having thus established the destitute life of the Goodyears through most of the years of Charles's experiments, Webster turned slowly and dramatically to face Horace Day. Here sat a man who had made a small fortune from shirred goods; Day, who lived in an elegant townhouse in New York City, with servants. He spoke mockingly of the "powerful sympathy exerted by Mr. Choate" on Day's behalf, "for the unhappy condition, the melancholy and depressed circumstances into which Mr. Day has been brought by his unfortunate and miserable connection with Mr. Goodyear."

Webster said, "Now all these touching observations would, I think, be very effective but for one thing. The counsel have not thought it worth while to offer a particle of evidence to sustain these assertions; not one particle."

Webster knew the importance of isolating Day as the one odious villain. He surgically separated Day from his attorneys: "I am sorry to say the defendant went altogether out of his way in his answer to say that Goodyear, having failed in his business as a hardware merchant, threw himself into the rubber business as a sort of refuge. What particular weight that has in its bearings on this case, I have not yet learned. I am happy to observe that this odious statement in the answer did not receive notice from my brethren of the bar—the opposing counsel; they would not defile their mouths by uttering it."

He went easy on Hayward, who early on had joined forces with Day to discredit Goodyear's claim on vulcanization: "There was a time when there was a disagreement between Goodyear and Hayward, and Mr. Hayward was foolish enough to set up some pretenses of his own, but was soon ashamed of it, and his chief merit is that he had the manliness to disclaim it." Then he closed in to further isolate Day: "I know but one man in the world who denies the originality of this invention. No one out of this courtroom denies it." He looked around deliberately at the sea of faces, then zeroed in on Day: "I know but one *within* the court who denies it. . . . It is Horace H. Day, the defendant in this case."

And what of Richard Collins, the miracle man of Baltimore, who, the defense had suggested, had spontaneously invented vulcanized rubber before Goodyear did? Webster merely shook his head, as if to imply that a response was hardly worth the effort: "I will not do injustice to any man. I know nothing of Mr. Collins. It is not my habit to deal harshly with witnesses. I will not call names; but I will say that Mr. Collins' testimony in this case appears to me to be totally incredible—entirely incredible. I have never, in thirty years' practice at the bar, heard or read any so improbable as his."

Finally, Webster attacked a contention by Day and Choate that the issue of damages could not be decided until it had first been ruled, by a jury, that Day in fact infringed on Goodyear. Webster dismissed this argument by pointing out Day had essentially admitted to as much when he settled with Goodyear in 1846: "I say it is unheard of, that a party charged with the infringement of a patent right, having had one fair chance to try that question in the presence of a jury, having voluntarily struck his colors in the presence of a jury, and suffered a verdict and the judgment to go against him by consent is ever entitled, on the face of the earth, to have another trial."

He was preparing to continue on some point when one of the judges raised his hand, saying, "Mr. Webster, no more proof is needed in that matter."

Webster said, "Then we excuse the honorable judge from any further hearing today."

On this subdued note, Webster bowed slightly to the judges. It had been a remarkable performance—of physical stamina, of eloquence, and of Webster's phenomenal ability to understand and synthesize into a coherent argument countless minute details on a subject he had known little or nothing about just a few weeks earlier. Webster had not wished to take this case. But once he agreed, he had given the Goodyear forces every bit of effort he had. With a victory bonus of six thousand dollars in the balance, Webster had spoken for two solid days, flawlessly weaving specifics on the history and science of rubber with rhetorical flourishes on the rights of man. When the *New York Daily Tribune* published the entire speech a couple of months later, it occupied fourteen columns and nearly three large pages of tiny, eye-straining type.

The packed courtroom maintained a respectful silence as the old lion turned and walked deliberately back to his seat, exhaustion plainly showing on his face. Many in the audience must have realized they were witnessing not just the end of a summation speech, but the end of an era. It was the last time Daniel Webster would ever argue a case in court.

Webster left for Washington by coach that evening. Goodyear and the rubber manufacturers convened at Goodyear's hotel. Though the decision would not be handed down for several months, Goodyear and his licensees were in high spirits. Yet while others hashed out the finer points of Webster's speech, picked apart Day's case, or speculated on the judges' decision, Goodyear's attention quickly shifted back to his favorite subject, his one consuming passion: rubber. For Goodyear, the trial had been an extended irritation that kept him from his experiments. Through the course of the trial, he had filled his hotel room with a gallery of soft and hard objects: combs, knife handles, canes, hats. He excitedly passed them around to his visitors. Goodyear was proudest of a soft rubber globe, two or three feet in diameter, with the countries of the world printed on it.

Later, the participants gathered in the hotel dining room for a banquet. L. Otto P. Meyer, a young rubber manufacturer from Germany, attended the last day of Webster's speech, and the banquet. He recalled a half century later, "Nearly all the principal men interested in the Goodyear patent sat at the *table d'hote* with him, at immense tables, my friend Conrad Poppenhausen and I sitting together at one. Joy reigned all around, tuned by champagne. I wonder if many are yet living who were so happy at that feast."

That June, Goodyear returned to England. He was still out to prove to the world that he was the original inventor of vulcanization. While he waited for word on his battle with his American nemesis, he turned his attention for a time to his English one. It's not clear whether Goodyear and Hancock ever met face to face. Neither of them recorded such a meeting. But there's tantalizing evidence that they did. Goodyear's handwritten address book, which survives under a glass case at the Goodyear Tire & Rubber Company in Akron, contains Hancock's Stoke Newington address. At any rate, agents for Goodyear and Hancock did meet to try to work out a compromise that would allow vulcanized goods from America into England without a court battle. But the negotiations collapsed, and the issue would not be resolved until the case in Queens Court four years later.

A few weeks after the Trenton trial, Daniel Webster left Washington for a rest at his beloved estate in Marshfield. On May 8, a cool, sunny Saturday morning, he set out with his clerk, Charles Lanham, on a fishing trip to a pond in Plymouth, ten or twelve miles from his home. They rode in a high phaeton carriage driven by two horses. About nine miles into the trip, on a long, uphill stretch, a bolt connecting the phaeton's front and rear axles snapped. The carriage slammed to the ground, tossing Lanham and Webster headlong over the front. Webster broke most of the fall with his hands and arms, but his head struck the ground in what at first seemed a glancing blow.

Lanham, unhurt, helped Webster to a nearby cottage, whose awestruck owner could barely speak, knowing he had the godlike Daniel Webster recuperating in his bed. After a day of rest, Webster felt well enough to move. He believed at first that he had escaped with a minor forehead injury and swollen wrists and arms. But in the coming weeks his arms grew increasingly painful. His left arm, discolored from wrist to shoulder, had to be kept in a sling. More ominously, his behavior became increasingly strange. Servants noticed him wandering about the house, peering repeatedly into trunks and closets. In public, he seemed confused and unsteady. The modern medical diagnosis is that Webster suffered a subdural hematoma—a swelling in the membranous layers of the brain. The effects of the fall, combined with his worsening cirrhosis of the liver from years of heavy drinking, left Webster bedridden by the end of the summer. Weak, feeble, barely able to sit up, and suffering from a variety of ailments ranging from the neurological repercussions from his fall to catarrh and inflammation of the bowels and stomach, the high-living man of the world, famous for his love of fine food and drink, was reduced to a diet of milk mixed with limewater, gruel, and a thin soup, consumed in his bed. He continued to write optimistic letters to friends about his recovery.

Webster was still bedridden when, in late September, word came down from Trenton that the judges in the Great India Rubber Case had reached a decision. It was a complete victory for Webster in his last great court case, a vindication of Goodyear's patent, and an unequivocal slam on Horace H. Day. In separate statements, Judges Grier and Dickerson dismantled Day's claims point by point. Of the two, Grier's ruling was the more forceful and impassioned, citing the gift that Goodyear had given to the world, the sacrifices he had endured, and the bitter reward he had reaped.

"He spent his money, his time, his credit," Grier wrote. "His family was reduced to poverty, and himself immured in a prison. Yet still he persisted. And finally his faith and energy were rewarded by the discovery, the value of which this controversy most abundantly proves. As we know a substance only by its qualities, he may be said to have discovered a new substance."

Without mentioning Day by name, Grier delivered this broadside against a man he clearly considered a skunk: "And yet when genius and patient perseverance have at length succeeded, in spite of sneers and scoffs, in perfecting some valuable invention or discovery, how seldom is it followed by reward! Envy robs him of the honor, while speculators, swindlers, and pirates, rob him of the profits."

Both justices upheld the validity of Goodyear's patent and, more to the point, affirmed that Day himself had acknowledged its validity when he signed the 1846 agreement that landed him the shirred-goods license. The agreement had stated that Goodyear "has, owns, and controls all and every right and privilege, granted or intended to be granted to him, the said Charles Goodyear, by the several Letters of Patent issued to him."

Grier and Dickerson found the claims of the other "inventors" absurd. "Every unsuccessful experimenter who did, or did not, come very near making the discovery, now claims it. Every one who can invent an improvement, or vary its form, claims a right to pirate the original discovery," Grier wrote. "The testimony about Collins, which has been much relied on in the argument, contains much internal evidence of falsehood." Even assuming Collins had told the truth (and Grier felt he had lied), the judge would not have altered his opinion: "He may possibly have been so near the discovery as to stumble over it. Yet he never went on to perfect his experiments. They were abandoned without a successful result. Every factory that employed him, and purchased his services or his secret, failed. No goods of 'vulcanized India Rubber' made by him, or in the factories that employed him, have survived to prove his success. Time which devours all things, may be presumed not yet to have digested or destroyed all the vulcanized rubber supposed to have been made (if it ever had been made) by Collins and the various other pretenders."

Dickerson found Collins's claims "incredible for several reasons. In the first place, it cannot be believed that a young man, without any previous knowledge of the India-Rubber business, without ever having used, or heard of the use, of the different articles necessary to produce that material, should have discovered, at the first effort, those several articles, with the proportions of each, and the degree of heat necessary to produce the desired result. And it is equally incredible that, he having made so important an improvement, should not have availed himself of the benefit of that improvement during the twelve or fourteen years that he was engaged in the different rubber establishments in Boston, Lowell, Salem, & c., where every exertion was making to improve the manufacture of rubber."

Dickerson was especially scathing on Day's own claims to being the inventor of vulcanization, especially given the testimony of Horace Cutler, Goodyear's Springfield partner who had betrayed the vulcanization secrets to Day. Day's claim "is not confirmed by any witness on the part of the Defendant; but it is quite manifest from the letter of the defendant himself, explained and confirmed by the testimony of Horace Cutler, that until after December, 1842, he, the Defendant, did not understand the process of vulcanizing rubber, and that he then obtained the information from said Cutler."

In judgments that together comprised some twenty thousand words, there was scarcely a sentence or clause in which Day or his attorneys could find solace. Dickerson concluded his decision by saying of Goodyear, "I am entirely satisfied that he is the original inventor of the process of vulcanizing rubber . . . and that he is not only entitled to the relief which he asks, but to all the merits and benefits of that discovery."

The judges ordered a perpetual injunction against Day, preventing him from infringing on Goodyear's patents. The decision further instructed James S. Green, a master of the court, to take an accounting of all the damages Goodyear had suffered. Green was further instructed to account for all of Day's receipts on shirred goods, produced under

Goodyear's patents, and that Day be forced to pay any shortfall. Finally, Green was instructed to check Day's books "from time to time" in the future. It is unclear just how much money Day eventually paid to the Goodyear Association, or how much if any ended up in Goodyear's hands. Whatever it was could not have been enough to satisfy Goodyear's needs, for he was blazing a trail through England and Europe, spending money faster than it came in. But whatever the amount, the decision hurt Day badly. The *New York Times* estimated that Day spent a half million dollars pursuing Goodyear in various court cases over the years. But if he retreated to lick his wounds for a while, Horace Day was not through tormenting Charles Goodyear.

The victory, especially the financial windfall, revived Webster's spirits. If he could find two or three similar cases, he remarked optimistically to a friend, he could die free from debt. But it was not to be. His recovery proved to be brief and illusory. He spent his final days writing polite responses to letters from his creditors, assuring them he was doing everything he could to pay them. And he would not get full satisfaction for his work on the Goodyear case. One of his last letters, on October 8, was to George Griswold, the Goodyear licensee who had promised an extra $1,000 in case of victory but so far had not come through. "I trust you have been satisfied, as well as the rest of us, with the Decree, which has been entered up by the court in the India Rubber Case at Trenton," he wrote. "And I hope, My Dear Sir, that you will not think me obtrusive, if, in connection with this subject, I call your attention to your very kind and confidential letter to me on the 18th of March last."

Before he could receive a response, Webster took a turn for the worse. He lingered for several days, in and out of consciousness, with his family gathered around him. On the night of October 23, his doctor sat at his bedside, reading aloud from Webster's favorite poem, Thomas Gray's "Elegy Written in a Country Churchyard."

Gray's melancholy words, penned 110 years earlier, might have been written for the moment:

The boast of heraldry, the pomp of power,
And all that beauty, all that wealth e'er gave,
Awaits alike the inevitable hour.
The paths of glory lead but to the grave.

As the doctor read, Webster drifted off to sleep. The doctor finished the poem, closed the book, looked tenderly at the dying man, and left the room. Webster slept peacefully for a few hours. Shortly after midnight he suddenly woke and cried, "I still live!" Then he collapsed, and he died two hours later. Ten thousand well-wishers crowded Marshfield for Webster's funeral. Another fifty thousand gathered in Boston for a memorial service. At Dartmouth, Rufus Choate gave a moving eulogy to his foe, his colleague, his hero. Newspapers and journals across the country lamented the loss of a giant.

13

GOODYEAR IN EUROPE

Suddenly everybody was making money on rubber, thanks to Charles Goodyear's perseverance. An industry all but dead just a decade earlier now thrived. Manufacturers sprang up across Europe and America with a vigor that outstripped that of the initial boom of the 1830s. These included legitimate licensees of patent holders and countless illicit companies who, less bombastic and combative than Horace Day, often infringed unimpeded. During the height of the 1830s boom, the entire world market for rubber shoes amounted to fewer than a half million pairs, most of them formed on crude lasts on-site amid the rubber trees of South America and then imported to Europe and the United States. By the early 1850s, manufacturers operating under Goodyear patents alone were cranking out fifteen thousand pairs a day, or around five million a year, in ever more varied and sophisticated styles. By 1879, an industry worth practically nothing when Goodyear chanced upon the life preserver produced more than twenty-five million dollars' worth of goods worldwide. Although clothing continued to be a staple, new uses flowered every day. Rubber was proving to be a miracle substance after all, the miracle substance everyone had hoped it would be back in the 1830s. The reality of a substance that was waterproof, airtight, nonconductive, stretchable, and malleable—and, moreover would not melt or freeze—fired people's imaginations. Air mattresses, conveyer belts, packing material for heavy machinery, and rims for carriage wheels all began appearing in the United States, driving a revolution in industry. In Goodyear's hometown of New Haven, William Pollard by the late 1850s advertised rubber as "A certain cure for leaky roofs!" "It is no new discovery or patented humbug," Pollard assured readers of the 1859 city directory. It could be applied equally well to cloth, tin, shingle, or metal roofs.

Amid all of this excitement, Charles Goodyear, having returned to England to be near his display at the Great Exhibition, began planning for an even more extravagant display at a coming world's fair in Paris. While the rest of the world made money on rubber, Goodyear continued to spend. He poured his energies into ever newer and more offbeat projects. He was also busy writing. In 1853, he published the first comprehensive book on the rubber industry. The book, in two volumes, carried the lengthy title *Gum-Elastic and Its Varieties, With a Detailed Account of Its Applications and Uses, and of the Discovery of Vulcanization*. Like Goodyear himself, the book is brilliant, clumsy, disorganized, and engrossing. Considering the conditions of turmoil—illness, financial troubles, lawsuits, the compulsion to continue his experiments—under which he produced the book, it is astonishing that he managed to complete the project—or, rather, nearly complete it. Proving himself the forgetful scientist, Goodyear published the book before he'd proofread it: there are charming lapses throughout, directions to see a certain drawing that does not appear, here and there a word simply missing, as though Goodyear intended to fill in the information during the editing process, then forgot or ran out of time or money. In at least one case, an entire paragraph goes missing.

The book, as Goodyear envisioned it, was to be a testament to the wonderful properties of vulcanized rubber—not just in the words but in the construction of the book itself. Several copies were printed with hard rubber bindings and covers. At least one copy was printed entirely on rubberized tissue. Goodyear firmly believed that rubber would supplant paper in books. It was one of his less prescient ideas. He assumed vulcanized rubber to be practically impervious to heat, cold, and time. But even vulcanized pages eventually stick together. In the end, Goodyear printed only a few copies of his book. The expense of binding with rubber materials was enormous. Instead of capitalizing on his budding fame with a conventional book that might have sold well and earned him and his family some money, *Gum-Elastic* proved to be yet another expense.

The first volume explores history, both Goodyear's and rubber's, events leading to his discovery, and offers descriptions of the vulcanization process, compounding methods, methods of manufacturing, and reflections on patent laws. The first sixty pages of his book are the least remarkable, since they amount basically to boilerplate history taken, often in lengthy chunks, from various encyclopedias, histories, and travel writings. Ironically, Goodyear never set foot in a rubber-producing country. His entire experience involved imports. He never saw the trees from which flowed his own life's meaning and purpose. He wrote, "The most remarkable quality of this gum, is its wonderful elasticity. In this consists the great difference between it and all other substances. It can be extended to eight times its ordinary length, without breaking it, when it will again resume its original form. There is probably no other inert substance, the properties of which excite in the human mind, when first called to examine it, an equal amount of curiosity, surprise, and admiration. Who can examine, and reflect upon this property of gum-elastic, without adoring the wisdom of the Creator."

Gum-Elastic captures our interest around page seventy, when the author stakes his "claims of the author as inventor"—among them his pioneering of vulcanization and his stubborn devotion when the rest of the world seemed to have abandoned the stuff for good. The style is painfully formal and polite, yet he makes his points, as with this salvo conceived to set him apart from Hancock: "Pains have also been taken by him [Goodyear] to credit to other persons the inventions which originated with them, even though they have been developed and brought to the notice of the public by the writer."

Goodyear states in dispassionate voice his devotion: "He has, with little reference to personal comfort or pecuniary advantage, applied himself constantly to the development of the subject for the period of fourteen years, without being diverted from the fixed purpose to complete the system of invention, as presented in this work, avoiding the temptations that often presented themselves, in the profits which might be derived from prosecuting the manufacture of many of the articles, and has made invention and the improvement of gum-elastic his business and profession."

Indeed, Goodyear makes his fiscal irresponsibility sound like a natural by-product of his mission; he is ashamed that he had to think of money at all. It disturbed him greatly that this holy mission of his had to be dragged down to the mundane and low level of court battles with the likes of Horace Day: "Having confined himself to these labors for so long a time, it would have been indeed grateful to the inventor if none of them need to have been made subjects of patents. It is repulsive to the feelings, that improvement relating to science and the arts, and especially those of a philanthropic nature, should be made subjects of money-making and litigation by being patented. The apology he has to offer for doing that which was repugnant to his feelings, is the unavoidable necessity of the case."

Goodyear saw his biography as the story of rubber; all else was secondary. Personal details remained sketchy; the details on rubber many.

He predicted correctly that rubber would usher in a great age of electricity because it was nonconductive and could be easily molded around wires: "This substance is one of the best non-conductors of electricity, and it is but reasonable to suppose that advantages may be derived from some of the fabrics in connection with electric machines, on account of this property. An anecdote is told of a professor, who, having highly charged an electric machine while wearing India rubber shoes, and standing on the wire, upon taking them off, and resuming his position, he was convinced of the their non-electric property, by being knocked down without them." Goodyear's description anticipates rubber boots and gloves worn by generations of electrical workers, as well as wire insulation, which made commercial electricity possible. Without vulcanization, electricity could never have been harnessed on the scale it was.

Goodyear reflects deeply on the role of the inventor in public life, a section that makes especially interesting reading in light of his own ordeals: "It is a mistaken idea with many, that the invention of an improvement consists in the first vague idea of it. It takes far more than that to entitle one to the merit of an invention, for, between the bare conception of an idea, and the demonstration of the practicability and utility of the thing conceived, there is almost always a vast amount of labor to be performed, time and money to be spent, and innumerable difficulties and prejudices to be encountered, before the work is accomplished." Goodyear adds, "It is worthy of remark, that the greatest discoveries usually afford their authors less remuneration than is obtained by others for trivial inventions. The more important an invention, and the more it interferes with previously existing modes of industry, the more are the public interested to dispute the claims, and infringe upon the rights of the inventor."

For all of his stilted language and third-person formality, Goodyear could sometimes attain a sort of literary grace with a simple, heartfelt truth: "It is often repeated that 'necessity is the mother of invention.' It may with equal truth be said, that inventors are the children of misfortune and want; probably no class of the community, in any country, receive a smaller compensation for their labors than do inventors."

The second volume, *The Applications and Uses of Vulcanized Gum-Elastic*, is an exhaustive rundown of all the products and potential products he envisioned. Of the two, the second volume is the longer, containing sketches of hundreds of products. And you get the feeling that the author was trying mightily to rein himself in. Goodyear could have written a shorter volume covering the uses for which rubber was *not* appropriate. After scouring his imagination he came up with only two inappropriate uses. First, Goodyear assures us that he "is not so infatuated with the subject as to recommend it as an article of food." The only other fault he finds with rubber is this: "It should not be worn next to the skin, nor should one sleep enveloped by it; such are not the legitimate uses of the article."

Life preservers had drawn Goodyear to rubber in the first place, and he never lost his interest in them. He envisioned not just conventional life jackets but a slew of other products. The world traveler could be equipped to save him- or herself from drowning by grabbing any number of handy belongings. "A common sized trunk of this sort, measuring three cubic feet, allowing thirty pounds for weight, has buouyancy sufficient to sustain one hundred and fifty pounds of baggage, so that they may serve not only all the purposes of a life preserver, but at the same time will carry safe a large amount of specie or other valuables." Imagine, paddling away from a sinking ship, clinging to a trunk with one's valuables stored safely inside! The beauty! But Goodyear was not finished: "A number of them lashed together will form a safe raft, or if lashed on the outside of a boat, will make it perfectly safe for a much greater number of persons than it would otherwise carry." Suitcases, too, could be made to double as life preservers. If just one traveling bag didn't make you feel secure, you could opt for the protection of a "double life-preserving traveling bag." If your disaster was not shipwreck but fire, you might see yourself climbing down a rubber fire-escape rope. The rope would have "a cross rope and handles, in order that after the rope is secured and passed from an adjoining building, persons may pass from one building to the other, or descend to the ground."

No fanfare attended Goodyear's privately published book, the object of so many months' work. There would be no reviews, no praising journal articles; after all, fewer than a dozen copies were ever printed. Like most of his endeavors, the book cost him money rather than earning him any. The work would remain in almost complete obscurity for nearly a century, until in 1937 a British publisher produced an edition.

Through all the years of poverty and frustration, Clarissa provided the one constant positive in Charles Goodyear's life. She bore the suffering of her poverty and the deaths of her children with stoic grace. She remained remarkably healthy and strong, and always seemed ready to nurse her perpetually infirm husband back to some semblance of health. As Goodyear's health steadily declined in the early 1850s, Clarissa remained at his side. When gout flared, preventing him from walking, Goodyear would lie on a sofa while Clarissa read him passages from the Bible. At times, Goodyear's various ailments combined to make him almost insensible, and at those times Clarissa would sit silently with him in his room, caring for him until he regained strength enough to carry on.

To any observer of the couple, Clarissa was the rock, the constant, the force somehow holding things together; Goodyear was the inconstant flame, by turns brilliant and feeble. It would not have surprised anyone, at any time, to learn that Charles Goodyear had suddenly died. But it was unthinkable that Charles would outlive Clarissa, or that Charles could survive for long without Clarissa's steadying hand. Yet that is precisely what occurred. Some time in the early spring of 1853 in London, Clarissa Goodyear took ill. She died on May 30, 1853, three months short of her forty-ninth birthday. Goodyear was devastated by the loss. It seemed that he had lost just about everything close to him. Somehow, he would have to find a way to carry on without her.

Goodyear had settled in rooms overlooking Leicester Square in central London, where he would live off and on throughout his seven-year European stay. To compensate for the loss of Clarissa, he poured himself into preparations for the world's fair in Paris to be held in late 1854. He wanted this show to outdo even his performance at the Great Exhibition of 1851 in London. In March he was confined to the Leicester Square apartment with a flare-up of gout, but by the fourteenth he felt well enough to take an afternoon carriage ride. Encouraged, he wrote an associate saying he planned to be in Paris within a week to work on plans for the exhibit. But a relapse kept him from leaving.

He would not arrive in Paris until July. Before then, on May 30, he married for the second time. Fanny Wardell was a twenty-year-old Englishwoman, almost thirty-four years younger than Goodyear. It is not difficult to imagine why Goodyear, lonely, ill, and needy, would have been attracted to Miss Wardell—she was a healthy, robust, attractive young woman with the energy to look after him. What is harder to fathom is what Miss Wardell saw in Goodyear. It could hardly have been physical attraction. He was, at fifty-three, more than twice her age. And he was an *old* fifty-three, sickly, sallow, and hobbled. He was bedridden much of the time, and when he felt well enough to move about he usually did so on crutches to relieve the pain from his gout. Indeed, he was using them when he and Fanny first met. And while he had by this time garnered a certain degree of fame, he never lived the life of a swell. At twenty, Fanny was fresh and just peering over the horizon of her youth toward life's possibilities. Goodyear's life was a closing door. Though he never lost his animation when it came to rubber, his spirit, like his face, was etched with an incalculable sadness, deepened by latent guilt, over the losses he had suffered. Fanny must have been a preternaturally sensitive soul. When the Reverend Bradford K. Peirce interviewed her shortly after Goodyear's death, she would say of her attraction, "I shall never forget the impression made on my mind by the deep melancholy in the first tones of his voice that ever fell upon my ear, as, leaning upon his crutches, he was first presented to me." During the next six years she would bear three children by Goodyear, only one of whom would survive infancy, and would be his traveling companion and nurse in places far away from her native soil. This large-hearted woman would minister to his many medical needs, consult doctors, patiently administer medications, and care for him until his death. If much of Goodyear's life was marked by poor decisions and bad luck, he was singularly wise and fortunate in the women who married him.

For the summer and early autumn of 1854 Goodyear shuttled between Paris hotels and Leicester Square. As the opening of the exhibit approached, he left Leicester Square for a rented house at 42 Avenue Gabriel in the Champs-Elysées (with a sort of charming provincialism, Goodyear spelled it "Champs Elissie"). The house would serve as his headquarters for the next two years. Goodyear signed an extended lease and soon began to fill the place with furniture made of hard rubber, or ebonite. The home became a curiosity and a showpiece for all of Goodyear's French acquaintances.

Goodyear gave little thought to cost in assembling the French exhibit. Money from U.S. licenses went straight toward expenses. He decided to grace the exhibit with portraits of himself and others painted on slabs of hard rubber. His chosen artist was George P. A. Healy, an American living in Paris at the time, a celebrated portraitist whose subjects included Lincoln, Longfellow, Henry Clay, William Tecumseh Sherman, John James Audubon, Andrew Jackson, and John Quincy Adams, along with a slew of European royalty. His favorite subject had been his close friend Daniel Webster, whom he painted several times. According to Healy's biographer, Marie De Mare, Goodyear and Healy met some years earlier at Webster's estate. Presumably Goodyear had visited there in hopes of convincing Webster to take on the case against Horace Day, or, after Webster had agreed, in order to brief him. Healy had been in Marshfield to paint one of his many portraits of his favorite subject. In the course of the meeting Goodyear, as usual, began talking up the properties of rubber, even remarking, for Healy's benefit, that portraits would one day be painted on slabs of rubber. Healy laughed and said that would be impossible.

Now, in 1855, Healy was busy in his Paris studio when Goodyear knocked at his door, unannounced. He bore a large, flat package containing several panels of hard rubber. "Here they are, and here I am," Goodyear said with a smile. "Will you paint my portrait?" Healy, greatly amused, immediately agreed. The resulting portrait shows an idealized Charles Goodyear. Instead of the sickly man who had hobbled into Healy's studio, we see Goodyear sitting upright, gazing serenely to his left. It is clearly Goodyear, but Goodyear as he might have looked twenty years earlier, or if time hadn't taken such a toll. Luxuriant, dark hair, flecked with gray, sweeps back from his broad forehead. His wide, dark eyes brood over a long, straight nose and full lips. His expression is one of quiet confidence, as if all of the travails of the past twenty years had been mere trifles on the way to assured triumph. He wears a formal white shirt and tie with a high collar, and a long, flowing burgundy robe. In his left hand, adorned with a pinkie ring, is a cream-colored pouch or a rolled-up piece of paper—perhaps a copy of his patent. Goodyear was so pleased with the portrait that he commissioned Healy to paint portraits on rubber of his son Charles Jr., and, working from past drawings, of Daniel Webster and Clarissa. The portraits would be hung proudly at Goodyear's French exhibition. Goodyear would die owing G. P. A. Healy seven thousand dollars that he no doubt had intended to pay as soon as he was able.

To help promote his exhibit at the French fair, Goodyear enlisted the help of Daniel Coit Gilman, a young Yale graduate from Norwich, Connecticut. Gilman, who would later become famous as the first president of Johns Hopkins University, had come to Europe a year after graduating from Yale to serve as an attaché to the U.S. legation in St. Petersburg, Russia. The position left him plenty of time to roam England and France, study in Germany, and engage in the usual youthful agonizing over his future, which he recorded in bittersweet letters to his older sister, Maria. Goodyear and Gilman, who had probably known each other from New Haven, hooked up in England in early 1854. Gilman was supplementing his low pay by writing articles from Europe for various newspapers and journals, a task he found tedious. Goodyear instead proposed that Gilman, who was fluent in French, act as a sort of promoter and press agent, writing and arranging the printing of pamphlets about the coming exhibit, and also promoting rubber in Russia.

Gilman leaped at the opportunity. He wrote excitedly to Maria, in April 1854, "Goodyear's proposal was 'providential' indeed. It has relieved me in part from the necessity of [journalism]." The relationship began pleasantly enough, with Goodyear's praising Gilman's writing and reminding him to keep track of his costs, sometimes enclosing cash or bank drafts for a few hundred francs. As early as October 1854, before Goodyear even left the Leicester Square apartment for France, Gilman was already worried that Goodyear's payments were less than reliable. Goodyear attributed the "inconvenience" to the fact that he had been delayed in coming to France and had not been able to see Gilman before Gilman left for St. Petersburg.

By January 1855, though, Goodyear was feeling a serious financial pinch from his overextended exhibit. Although Gilman's letters to Goodyear don't survive, it is clear from the tenor of Goodyear's responses that the young man was becoming alarmed. More than likely Gilman wasn't just worried about his own fees, but was on the hook for payments to various printers for pamphlets and notices he had written for Goodyear. Gilman wanted to settle accounts and get out of the relationship. Goodyear, bedridden with various ailments, responded at last with a check, plus what for him amounted to defensiveness and anger: "I intimated in my last [letter] some dissatisfaction on my part, after very many expressions of dissatisfaction by yourself," Goodyear wrote. Later in the same letter, he added, "I had no notice at what point of time your agency for me terminated, and felt that you could either come to me if so disposed, or communicate to me by letter that which related to your agency, before pressing with such earnestness, the settlement of your account. If I am wrong, I shall be happy to stand corrected. If right, I have no doubt you will satisfy me as relates to what you have omitted."

Within a few weeks, Goodyear and Gilman patched up their differences. Most likely, Goodyear juggled his books and found a way to pay Gilman what he owed, and the relationship continued until Gilman returned to the United States near the end of 1855. By July, Gilman was hard at work on another pamphlet for Goodyear, in French, advertising the extraordinary properties of rubber. "Your outline for the pamphlet is very good," Goodyear praised. A still leery Gilman had asked Goodyear how much he should spend on printing costs. Goodyear, with the casual optimism that always marked his discussions of money—especially when he was doling it out to make or to promote rubber—responded, "I may say in general terms, all that is necessary to do justice to the subject. Wait upon the most influential editors, and make them fair compensation for taking up the discussion to which you advise." On August 1, Goodyear told Gilman he would be willing to pay up to five thousand francs for the brochure—"I think not more at present." It's safe to say that even five thousand francs was a stretch.

Although the most inconvenient aspect of Goodyear's ill health was the gout, he was dealing with health problems that were more ominous. As usual, his delicate condition was made worse by overwork and his restless obsession to move ahead with his rubber plans. On August 14, 1854, with Goodyear and Fanny staying at a Paris hotel while he was working furiously to prepare for the exhibit, Goodyear awoke in terrible pain. It was a condition to which he was accustomed. But the inventor, "being much pressed with business issues got up and worked hard all day," Fanny recorded in a medical journal. He returned from his shop that evening looking especially weak and haggard. Severe pain in his throat made it impossible for him to speak. He awoke at three o'clock the following morning in great pain. He could not move from bed for a week. A Dr. Bigelow, summoned by Fanny, prescribed a "constant poultice with linseed heat sprinkled with laudanum," plus regular doses of two medicines that Fanny doesn't name. When Goodyear began to recover, the doctor ordered frequent baths at thirty degrees centigrade, Bordeaux wine with dinner, total abstinence from milk, and "tonic medicine." No sooner had Goodyear begun to feel better than he resumed a hectic schedule of work and travel. On September 1, while staying at Aix-la-Chapelle, "the fatigues of business, traveling, and sightseeing produced a relapse," and Goodyear was confined to bed for a day and forced to use crutches for three more. He was suffering from gout and also painful bouts of indigestion, for which a doctor prescribed a series of powders, plus a tablespoon of Madeira wine four to six times a day.

On September 10, Goodyear experienced a slight cough at bedtime. Given his panoply of other ailments, this would seem a mild complaint, except that the cough brought up, in Fanny's estimation, about two tablespoons of blood. His hands were damp and chilly. His whole body became cold and marked by pallor. A Dr. Velten diagnosed the problem as "a bursting of a blood vessel in the bronchial membrane," Fanny recorded in a medical diary. He prescribed lemonade, plus one tablespoon every hour of acidi hallera, aqua ceraforum, and syrup ceraforum. He was ordered to take no animal solid food, wine, or coffee for twenty-four hours after the bleeding stopped, and to refrain completely from talking. Knowing Goodyear's predilections, the latter prescription would have been the toughest to follow.

Fanny's medical diary is incomplete—it covers just a few pages in an otherwise blank notebook. Perhaps she purchased another book, or simply became overwhelmed by the prospect of recording every horrific detail of her husband's medical woes. But the few pages are enough to demonstrate the agony in which Goodyear lived.

Buried elsewhere in the notebook is a curious find—two or three pages of ledger notes in Goodyear's hand, detailing some of his expenses, and giving some indication of just how easily cash slipped through his fingers. Withdrawals on his account at a London bank, listed only as "cash at sundry times," came to nearly £5,700 in September 1853, nearly £3,000 in January 1854, £560 in February, and £2,600 in March. That was a total of nearly £12,000 over seven months. By April, he had a remaining balance of just over £181 pounds. More enlightening than the figures themselves is the fact that one should find them buried among the unused pages of his wife's medical diary. Little wonder, with these accounting methods, that Goodyear, no matter how much or little money he earned, was in a constant state of confusion over his finances, constantly living on the edge of ruin.

For Paris, Goodyear devised an India rubber exhibit featuring $50,000 worth of goods. About half of these were holdovers from his Vulcanite Court in England. The rest were brand-new items. There were objects great and small—sofas, tables, chairs, bedsteads, pillars, pearl-inlaid boxes, eyeglasses, inkstands, surgical instruments, walking sticks, brushes, combs. The exhibit was one of the highlights of the world's fair, and it became a point of pride to both Americans and the French—the French because many of the objects had been made by French manufacturers according to Goodyear's specifications. The exhibit attracted the eye of the emperor, Napoléon III, who visited and was impressed with both the objects and the inventor. Goodyear and the emperor struck up something of a friendship: Goodyear presented the empress with an ornate chatelaine made of hard rubber. Napoléon III invited Goodyear to take carriage rides with him through the streets of Paris. But while he brushed shoulders with nobility, Goodyear's financial troubles worsened. To pay for all of the new goods and their shipping, he needed a steady flow of cash from his far-flung licensees. When the licensees fell behind, or paid less than he expected, he was in a bind. He had signed off on thousands of dollars' worth of goods and machinery, which he planned to pay for with the proceeds from the French licensees. Now he was liable for significant payments that he had no practical way of meeting. The result proved to be depressingly familiar. Goodyear was resting at his rooms one evening in early December when two French officers knocked at his door. Fanny answered and let them in; Goodyear had just fallen asleep. It was reported that another officer guarded the balcony window, as if the hobbling inventor might attempt an escape. The officers took Goodyear down to Clichy, the Parisian debtors' prison, where he was locked up in his old familiar world, the debtors' cell. Fanny accompanied him to prison and asked to be permitted to stay with him in his cell. She was refused.

During the four days that Goodyear spent in the prison, he learned, ironically, that he had been awarded both the Grand Medal of Honor and the Cross of the Legion of Honor for his exhibit. He received the medals from his son Charles Jr. while sitting in jail.

14

FAME AND SADNESS

For all of his troubles, Goodyear, in the wake of the Trenton trial and the exhibitions, was emerging as a famous man. A Goodyear legend was on the rise. His own countrymen, who had so long dismissed him as a harmless quack, celebrated him in journals and newspapers. His story began to fit popular legends of the long, hard struggle followed by just reward. One of the more curious entries came from the *American Phrenological Journal*, a periodical devoted to a popular nineteenth-century fad science that sought to connect character traits with the size and shape of the skull: "The temperament of Mr. Goodyear is a combination of the mental and vital, or Nervous-Sanguine. Warmth, zeal, sprightliness, and restless energy, seem to be the leading characteristics of his physical constitution. Whatever he undertakes, he prosecutes as if it were worth doing promptly and well. His mind acts rapidly, but clearly; and though he seems changeable, he makes thorough work as he goes. A plodder would not have had time to make half as many experiments as he did; and a man who had not an almost instantaneous power of perception and analysis, would have derived from those experiments no valuable information."

An assessment of Goodyear's skull reveals that "the reader will observe a swelling-out of the upper part of the organ of Constructiveness, where it joins Ideality, at the temple, which indicates the tendency for invention, experiment and discovery. Rising above this toward the centre of the top of the head, a considerable elevation and fulness [*sic*] are seen." Later, we learn that "Hope, Firmness, and Combativeness are also large in his head. These he exhibited most signally in that energy which laughed at obstacles, in that firmness which would take no denial, and in that unquenchable confidence in ultimate success, which sustained him through so many discouraging circumstances, and bore him onward to final triumph."

In a final rhetorical flourish, the author showed that Goodyear's suffering was already transforming into a Goodyear myth: "His hope, undimmed, burned with unabated fervor in the darkest hour, and thus sustained him until his conquest was completed. We saw him, haggard and worn, and weary, in the darkest hours of his struggle, and, though he was pointed out to strangers as the man who was crazy on the subject of India rubber, we saw in the pale and care-worn man, the faith and hope that, though cast down, are not destroyed, and a gleaming fire in his eye that bespoke perfect confidence in himself and his great idea."

A somewhat more conventional assessment appeared in the mid-1850s in the *Illustrated American Biography*, which offered a picture of a man who had endured a long, hard struggle and was now enjoying the material rewards bestowed upon him by a grateful world. "Few inventions have done more to increase human comfort than the process by which caoutchouc, or India-rubber, is kept, in a perfectly pliable and soft condition, amid all the changes of the atmosphere. It is within memory of many now living, that India-rubber was used only to erase pencil marks from paper."

The article found that Goodyear had endured all of his hardships with "perfect cheerfulness." And some hardships they were: "Money, time, health, and all were wasted in the vain attempt, yet the stout heart of Mr. Goodyear never fainted. Disappointment only stimulated to further trial. His money was all gone, and credit soon followed. Then came lawsuits, duns, executions, sheriffs, and the sharp tooth of poverty. But nothing could daunt his invincible spirit, his indomitable courage." This succinct morality tale closes with the reassuring report that Goodyear had come through all of his hardships in fine fettle: "He is now in the full prime of life, and has already won a fame, throughout the world, equal to his deserts."

The man in the "prime of life" was in fact on his back much of the time, unable to walk, racked by dyspepsia and financial worries, coughing up blood, hounded by creditors and faced with the constant threat of jail. If the press bestowed on Goodyear a happy ending, life offered a more dubious finale. In France, Goodyear had reached the apex of both renown and humiliation. He had been awarded highest honors at the French world's fair and spent lonely nights in debtors' prison. He rode in carriages with the emperor, but could not keep his creditors away. He was becoming famous, but fame never provided a buffer between Charles Goodyear and hard times.

Released from Clichy Prison in Paris, Goodyear returned to England in early 1855, where he was promptly arrested and jailed briefly for a debt related to the French exhibit. It would be Goodyear's last encounter with debtors' prison, a little over twenty years after his first. As a mark of his growing renown, his surviving address book from this time contains, in addition to the expected roundup of rubber suppliers, chemical companies, and banks, the name of Charles Dickens, who was in his early forties and at the height of his fame. Dickens had recently published his dark masterpiece *David Copperfield*, based in part on his own experiences as a child accompanying his wastrel father to the debtors' prison. Dickens and Goodyear would have had much to talk about.

Another address listed in Goodyear's book was Thomas Hancock's Stoke Newington factory. At some point during Goodyear's time in

England, the two old foes may have met face to face for the first time, in an attempt to avoid a courtroom showdown between Hancock and Goodyear's English allies, led by Stephen Moulton. Goodyear wanted compensation for money he'd lost because of Hancock's patent, as well as the right to sell his products in England. If the meeting or meetings occurred, they at any rate ended without an agreement. That summer, the long, slow legal battle finally saw its resolution, in a highly public forum before the Court of Queen's Bench, Guildhall, in July 1855, just two years before Hancock's patent was due to expire.

The case was similar to Hancock's suit against the British shoe importers several years earlier, with the important difference that this time Hancock was the one on the defensive. Moulton had convinced the crown to examine Hancock's patent under a provision known as scire facias, by which he hoped to have the patent declared invalid. The provision essentially required Hancock to prove the validity of his patent. The two-day trial was a publicly humiliating ordeal for Hancock, from start to finish. Moulton took the witness stand and recalled his journey bearing Goodyear's samples, and his conversations with Brockedon and the Birleys. Next, Hancock was forced to watch as Goodyear hobbled to the witness stand and recounted his excruciating odyssey with rubber, the setbacks, the poverty, the breakthrough in Woburn, and his unrelenting determination. As Horace Day had learned, Goodyear's very appearance—the too-large eyes, the sunken cheeks, the deep worry lines—made a powerful case.

Next, Hancock faced searing questions from the prosecutor, H. Hill, who openly questioned Hancock's ethics. With Goodyear sitting quietly in the gallery, Hancock was forced once again to admit that he had examined the samples from America before embarking on his own quest to vulcanize rubber. Hill argued that Hancock had not, in fact, invented vulcanization at all, but merely copied Goodyear's methods.

Further, he argued that Hancock had not understood the process when he first applied for his patent.

But for all the squirming Hancock endured, the ultimate verdict was never really in doubt. As in the shoe case, Hancock's opponents faced almost insurmountable odds. The law was not concerned with who originated the idea, merely who made it to the patent office first. The jury found for Hancock. He declared a complete victory, implying in his memoirs that the verdict amounted to validation of his claims as inventor. But this was not so. Hancock did not mention that Lord Campbell, presiding over the case, made a point of publicly rebuking Hancock's behavior: "If Goodyear's invention was prior in point of time, it was not handsome in Hancock to look at his specimens and try to find out his discovery; and if Goodyear was the inventor, it was to be regretted that he should not have the benefit of his invention." Such words, from so high a source and so publicly delivered, haunted Hancock. His fortune was secure, but his reputation had taken a blow from which it would never fully recover.

When he felt well enough, Goodyear continued his experiments with rubber at 47 Leicester Square. His greatest work was already behind him, but he continued furiously. Charles Jr. acted as his assistant. Although Goodyear had never worked out a formal business partnership with Moulton, the two stayed in close contact. Goodyear lacked the machinery and manpower for large-scale preparations. So he peppered Moulton with orders for quantities of rubber, often prepared to highly specific instructions. One letter asks Moulton to "send us 100 lbs of best Para rubber with 36 lbs of finest sulph. Mixed as before. One half of this quantity on cambric ¼ inch thick as free from air as possible." Another letter requests "100 lbs best Para mixed with 35 lbs sulphur run as above full ⅛ in thick. We prefer that the surface *not* be chalked." Goodyear was also working with gutta-percha: "We have sent you today some more gutta percha. Please mix about 10 pounds of it with coloring matter to give it a pinkish tint—like the item enclosed."

The constant barrage irritated Moulton, who was at the time trying to establish his own line of rubber products at his Bradford mills. When Hancock's patent finally expired in 1857, Moulton was busy manufacturing coats, capes, leggings, overalls, pantaloons, blankets, horse covers, wagon covers, sheet rubber for bandages, water tanks, and rubber suction hose in various widths. But Moulton's loyalty to Goodyear overrode his annoyance; he continued to fill Goodyear's orders faithfully.

Goodyear began making plans to move from London to Bath, a resort town about a hundred miles to the west. The mineral springs that give Bath its name, gushing a quarter of a million gallons of steamy water from the earth's core each day, had been cherished for their healing properties since pre-Roman times. Fanny hoped the move would reverse her husband's ominous slide. For Goodyear there was another attraction: Bath lay just nine miles from Moulton's factory in Bradford.

Goodyear's planned move in mid-February 1857 was delayed by illness. Charles Jr. wrote to Moulton on February 18 to say that his father had been "*dangerously* ill," but was now feeling better. Nevertheless, Goodyear was unable to complete the journey for two more months. In mid-April, the Goodyears arrived at last in Bath, where they rented a house at the lower end of Gay Street, a prominent thoroughfare of Georgian townhouses. Gay Street was built during a spectacular eighteenth-century renaissance that gave the city much of its current architecture. The street leads from Queen Square up a hill to the Circus, a stately array of nearly identical townhouses arranged in a perfect circle around a circular park. A half century earlier, Jane Austen, who wrote about Bath in *Northanger Abbey* and other works, had occupied a house on Gay Street, just across the street and up a few doors from the one Goodyear rented.

From his new quarters Goodyear continued to order supplies from

Moulton. "Father says he has ordered two hundred lbs of Para gum with 2 oun sulphur to the lbs," Charles Jr. wrote on May 16, along with this complaint: "The last thin sheet which you sent was much too thin for the purpose I wished." Goodyear, when he felt up to it, made repeated visits to Moulton in Bradford. For all his ill health and business worries, he showed characteristic kindness when one of Moulton's dogs followed him back to Bath. He wrote in a shaky hand, "A dog which I think is yours followed our carriage and as we could not drive him back we have put a strap on his neck and will try to get him home safe when I come to Bradford Monday or Tuesday."

The Goodyears would remain in Bath for a little more than a year before returning to the United States in May 1858. Goodyear owed Stephen Moulton money. An associate of Goodyear's, remaining in England, wrote to Moulton to ask for his continued patience: "Under the circumstances, I think we ought to make some allowance for Mr. Goodyear, who is doubtless overburdened with business cares, at present, without yet realizing any of their benefit."

Hancock, now past his seventieth birthday, published his memoirs in December 1857. The full title of the volume is a mouthful: *Personal Narrative of the Origin and Progress of the Caoutchouc or India-Rubber Manufacture in England.* Hancock clearly intended the volume to cement his position as the father of rubber. But the book is much more forceful in its claims regarding the early inventions, such as the masticator, than on vulcanization. His reticence on vulcanization is instructive, since it is clearly the most important development in the history of rubber. The section on vulcanization reads as though it was agonizing for Hancock to write, as indeed it must have been. Having already acknowledged in court that he had seen the samples from America before producing his own, Hancock could hardly now claim otherwise. Instead, in his memoirs he downplayed the importance of the scraps by emphasizing their size. They were "small bits," "small pieces," "very diminutive bits." Hancock's memoirs are not the work of an outright liar—there are no bald-faced whoppers to expose. It is rather in what Hancock neglects to say that one may find fault. Even at this late date he simply could not bring himself to properly credit Goodyear's achievement, and his failure to do so reflects poorly on him. Not once in the book's 283 pages does Hancock mention the name Charles Goodyear. As one British historian said with delightful brevity: "Hancock was economical with the truth."

When the book appeared, Hancock mailed gift copies to a number of people, including some of the leading scientists of the day. Michael Faraday, to whom Hancock had supplied latex samples to determine the molecular composition of rubber, sent a cordial note back, promising to read it, "when I can find time—I have not much of that at liberty just now."

Hancock also sent a copy of his memoirs to Nathaniel Hayward in the United States, along with a letter, pointed in its failure to mention Goodyear and seemingly calculated to inflame Hayward's latent resentment: "Considering the position you are entitled to take amongst the foremost in the manufacture of India Rubber in the United States I have thought that having written a simple narrative of the part which I have taken in its manufacture in this country a copy would not prove unacceptable to you. . . . If you have not already done the same thing in America no man I should think is more able or entitled to fulfill such a task." Hancock added that he hoped Hayward lived long enough to see sufficient credit given to him.

When Goodyear returned to the United States in May 1858, it was to seek an extension on his U.S. patent, nearing expiration. Goodyear wrote to Moulton explaining that his U.S. licensees "urged me to return by the first boat." To win the extension, Goodyear and his licensees would have to surmount the combined objections of Goodyear's two principal foes, Thomas Hancock and Horace Day.

Throughout his career, Hancock had concentrated exclusively on the rich commercial markets of his native Great Britain, never seeking inroads in the United States. Yet now, in 1858, past seventy years old, he traveled to Washington to testify against Goodyear's efforts to extend his U.S. vulcanization patent. Hancock's decision was motivated at least in part by the opportunity to open growing U.S. markets to Charles Macintosh & Company goods. An end to Goodyear's patent would make it easier for them to manufacture exports. They could import raw rubber into England, manufacture goods, then ship them to the United States, without ever paying import duty on the raw rubber, since the finished goods would be meant solely for export. Standing in the way of this neat plan was Charles Goodyear. But there was almost as certainly a personal reason as well for a man of his age to make such a long voyage. Since Goodyear had testified against Hancock in 1855, in the trial where Hancock faced the humiliation of Lord Campbell's "not handsome" comment, here was Hancock's chance to pay him back on Goodyear's home turf.

Day, though officially banned from any rubber manufacturing other than shirred goods, had not stopped making a nuisance of himself. Goodyear's brother Nelson, in 1851, had taken out a U.S. patent for hard rubber, made by increasing the concentration of sulfur in the vulcanization process. Through the mid-1850s, holders of licenses under Nelson Goodyear complained that Horace Day was infringing on the patent. Day, despite his courtroom setbacks, scoffed at the charges. What is clear, though, is that Horace Day had managed to become a wealthy man thanks to Goodyear's genius and perseverance. At various times, Day owned factories in New York City, Philadelphia, New Jersey, and Massachusetts. His personal fortune was estimated by the *New York Tribune* to reach a million dollars at least twice during his lifetime. By the time Goodyear returned to the United States a physically and financially broken man, Day was living in comfort in Manhattan and using his fortune to become a man of affairs. An enthusiastic supporter of the Republican Party, he befriended Nathaniel Prentiss Banks, the youthful governor of Massachusetts, and for several years was a principal donor to the *Atlas & Daily Bee*, a Republican newspaper. In 1859 he divorced his second wife, Catharine. A year later, city records show him living in a three-story, one-hundred-thousand-dollar house on Broadway with his third wife, Sarah, a Rhode Island native who was thirty-five years old, ten years Day's junior. Taking care of their personal needs were a coachman named Henry Simmons, a thirty-year-old Irish servant named Bridget McMehen, and two additional servants, Ann Jefferson and Henry Young.

The three-story house gave Day plenty of space to pursue another of his passions, Spiritualism. For several years, Day was the publisher of a journal called the *Christian Spiritualist*, which operated its headquarters on the third floor. Spiritualism, based on the proposition that one might, if properly attuned, commune with the dead, was all the rage in mid-nineteenth-century United States and Great Britain. In March 1848, the Fox family of rural Hydesville, New York, near Rochester, began to hear inexplicable knocking noises in their farmhouse at night. "The first night we heard the rapping we all got up, lit a candle, and searched all over the house," Mrs. Fox later remembered. "The noise continued while we were hunting, and was heard near the same place all the time. It was not very loud, yet it produced a jar of the bedsteads and chairs, that could be felt by placing our hands on the chairs, or while we were in bed." The noises cost the Fox family several nights' sleep. Finally, on the evening of March 31, the family went to bed earlier than usual, just before sunset. The knocks came again, louder and more insistent than usual. Finally, two daughters, Katie and Margaret, began snapping their fingers to mimic the rapping. Soon, the rapping was responding to the snaps. As fast as the girls snapped, the responses came. Eventually, the terrified but intrigued family was asking the knocker specific questions. The mysterious intruder rapped out the daughters' ages, twelve for Katie, fifteen for Margaret. Over the next several hours, during which time neighbors were called in who corroborated the story, the knocker revealed himself to be the restless spirit of a thirty-one-year-old peddler claiming to have been murdered by a former tenant and buried under the house.

Word of the strange happenings spread rapidly. Soon, previous tenants of the house came forward, claiming that they, too, had heard rapping noises in the night. In Stratford, Connecticut, a Reverend Eliakim Phelps reported that furniture in his house had been rearranged by other than human hands. Bread, sugar bowl, eggs, and other kitchen items were found scattered around the house. In one of the bedrooms, a nightgown was found under a sheet, with arms folded over the chest like a corpse's before burial. Strange, indecipherable writings were scribbled on the walls. At first the family thought they were the victim of some vandals' practical jokes, but when objects began flying about the rooms it became clear to them that otherworldly forces were at work.

The tales captured the attention of a nation already dabbling in phrenology, mesmerism, and other trendy pseudo sciences and religions. In contrast to the leader-directed worshipping in a conventional church, Spiritualism offered the opportunity for individuals to take control of their own religion. Indeed, established churches were among the harshest critics of spiritualism. What need was there for a hierarchical church and church-appointed men of God, when teenage girls in upstate New York farmhouses could commune with the spirit world?

Suddenly, just about anyone could tap into the spirit world, assuming one either had the calling or had access to someone with the calling. Every town, no matter how small, had at least one professional or amateur medium. A hundred and twenty years before the writer Tom Wolfe would put the label on the 1970s, the 1850s emerged as America's first real "Me Decade." There were many high-profile critics of Spiritualism, to be sure, but adherents were by no means relegated to the lunatic fringe. Prominent writers, politicians, physicians, and businesspeople on both sides of the Atlantic counted themselves among the believers, or at least dabblers. Nathaniel Hawthorne attended séances, as did Elizabeth Barrett Browning and Sherlock Holmes creator Arthur Conan Doyle in England. Mary Lincoln, the first lady, became obsessed with Spiritualism, holding séances in the White House, after the death of her beloved son, Willy. The president, though more skeptical, is known to have attended séances.

Horace Day embraced a branch of Spiritualism that attempted to reconcile itself with Christianity, hence the title of his magazine, *The Christian Spiritualist*. The third floor of his house served not only as editorial offices, but also as an unofficial gathering place for believers and mediums. The Fox sisters, who had become international celebrities, giving séances before crowds of several hundred, were frequent guests at Day's home. Day corresponded with Sarah Helen Whitman, one of the leading proponents of spiritualism of the era. Whitman, who wrote poetry inspired, she said, by contacts with the spirit world, was the former fiancée of Edgar Allan Poe. When the Society for the Diffusion of Spiritual Knowledge was formed, with a former Wisconsin governor as its president, Horace Day was named vice president and trustee.

Now in his late forties, Day retained his strong, athletic build. His full head of hair was turning white. Whatever his communion with spirits told him of otherworldly conduct, in the material world Horace Day had lost none of his bumptious combativeness. He still had it in for Charles Goodyear.

The Goodyears landed in New York in the late summer of 1858 and settled in New Haven, in a house at 211 Chapel Street. Goodyear began planning his case to extend his patent. This would be in many ways a tougher case to win than the Trenton case against Day. The patent office, then as now, operated under a goal of turning inventions over to the public domain as quickly as possible once the inventor had realized a fair return on his invention. With the sentiment of the patent office by design leaning in that direction, it falls to the applicant to argue why he has not yet made enough money. The essential argument comes down to the desire to benefit personally from one's invention versus the need to satisfy a cause at once as irreproachable-sounding as it is nebulous, "the public good." In Goodyear's case, proving financial hardship would not be easy, even if he was, in fact, poor. According to his own records, tabulated with the help of his attorneys, Goodyear had received $162,894.09 from licensees since having first been issued the patent. It was a good deal of money out of which to extract a claim of poverty.

"The originality of this patent has never been passed upon by any jury, since it was first issued in 1844, up to this time, in the United States," Day wrote in his challenge. In addition to his attempts to invalidate Goodyear's patent, Day suggested that Goodyear had been sloppy in using the invention, and had been less than diligent in seeing it turned into products. Therefore, the public good demanded that the patent be made public.

Public sentiment would not lean in Goodyear's favor as easily this time as last. As the biographical sketches quoted earlier in this chapter demonstrate, there was a widespread feeling afoot that Goodyear had "made it" financially—that he was a rich man. For one thing, every item that a licensee produced was stamped with the Goodyear name. Although his royalty payments on any given item were usually absurdly small, in the public mind Goodyear was selling all of these goods himself, and hence must have been rolling in cash. In fact, the nearly $163,000 Goodyear had received from his licensees was balanced by $130,000 in expenses, leaving him with a balance of $33,000 in profits. But his list of expenditures did not even begin to cover all of the people to whom he owed money. He was in hock to his ever patient brother-in-law, William DeForest, for more than $45,000 alone.

It may well have helped Goodyear's case that Horace Day was thoroughly disliked within the patent office. As early as 1851, a year before the Trenton case, Day had engaged in a nasty and highly public feud with Patent Commissioner Thomas Ewbank and a patent examiner named Dr. Gale over the original decision to rewrite Goodyear's patent in 1849. Day had also disputed Edwin Chaffee's right to a patent for his landmark rolling machine, the Monster. Day publicly charged Ewbank and Gale with favoritism, and claimed that Ewbank failed to read testimony before making crucial decisions. In a vitriolic, eighteen-page "public letter," Day asked, "Are such men fit to preside over the bureau, which has in its charge the sacred rights of inventors and the public? I leave it for the public to decide." By the time of Goodyear's patent hearing in 1858, Day's enemy, Ewbank, had left the patent office and been replaced by Joseph Holt. But it's safe to say that Horace Day had few friends inside the U.S. Patent Office.

As Goodyear sat in Washington watching the hearings, his confidence grew. On June 18, 1858, he wrote to Moulton in England to report that "the prospects of the extension of my patent seem to be very fair indeed." His confidence was borne out when Joseph Holt released his decision, an unequivocal affirmation of Goodyear and a blistering rebuke of his enemies, mainly Horace Day, whom he did not name. Holt's decision was, if anything, even stronger than had been the decisions by Dickerson and Grier in the Great India Rubber Case six years earlier.

"No inventor probably has ever been so harassed, so trampled upon, so plundered by that sordid and licentious class of infringers known in the parlance of the world, with no exaggeration of phrase, as 'pirates,'" Holt wrote. "The spoliations of their incessant guerrilla warfare upon his defenseless rights have unquestionably amounted to millions. In the very front rank of this predatory band stands one who sustains in this case the double and most convenient character of contestant and witness; and it is but a subdued expression of my estimate of the deposition he has lodged, to say that this Parthian shaft—the last that he could hurl at an invention which he has so long and remorselessly pursued—is a fitting finale to that career which the public justice of the country has so signally rebuked."

Holt took particular umbrage at Day's (fairly apt) suggestion that Goodyear had been lax in seeking useful products for his invention. Goodyear "seems to have been incapable of thought or action upon any other subject," Holt wrote. "He had no other occupation, was inspired by no other hope, cherished no other ambition. He carried continually about his person a piece of India-rubber, and into the ears of all who would listen he poured incessantly the story of his experiments and the glowing language of his prophesies. He was . . . completely absorbed by it, both by day and night, pursuing it with . . . almost superhuman perseverance. Not only were the powers of his mind and body thus ardently devoted to the invention and its introduction into use, but every dollar he possessed or could command through the resources of his credit, or the influence of his friendship, was uncalculatingly cast into that seething cauldron of experiment."

Holt added, "The very bed on which his wife slept, and the linen that covered his table, were seized and sold to pay his board. . . . His family had to endure privations almost surpassing belief. . . . We often find him arrested and incarcerated in the debtor's prison, but even amid its gloom his vision of the future never grew dim, his faith in his ultimate triumph never faltered."

Holt even forgave Goodyear his atrocious bookkeeping, which attorneys for Day and Hancock had hit especially hard. Goodyear's books "have not the precision and symmetry which belong to the products of the counting room, and which might have been imparted to them by the applicant, had he been a merchant's clerk instead of the brilliant and impulsive genius that he is," the commissioner conceded. But he added, "Inventors . . . have ever been distinguished for a total want of what is called 'business habits.' Completely engrossed by some favorite theory, they scorn the counsels and restraints of worldly thrift, and fling from them the petty cares of the mere man of commerce as the lion shakes a stinging insect from its mane."

It must have been particularly galling to Day and Hancock, both meticulous when it came to matters of finance, to see Goodyear's sloppiness thus held up as an example of his genius. Holt even gave Goodyear the benefit of the doubt when it came to his claims of profits on his balance sheets: "It is probable, indeed, in the view of the whole testimony, it is my firm conviction, that, if it were possible to extract from the tangled mazes of the multifarious and now half-forgotten transactions connected with the invention, all the moneys expended therein, it would be found that, instead of there being a balance to its credit, the balance would be on the other side." As a final blow to Day, Holt found that despite Goodyear's unchecked profligacy when it came to his rubber expenditures, he was in the rest of his life "temperate, frugal, and in all respects, self-denying."

Despite the presentiment to deny patent extensions, Holt found that "I should be false to the generous spirit of the patent laws, and forgetful of the exalted ends which it must ever be the crowning glory of those laws to accomplish. The patent will, therefore, be extended for seven years from the 15th of June, 1858."

For Goodyear, the victory was perhaps sweeter even than the court case from 1852, since it involved not just the pernicious Day, but Hancock as well. Hancock and one of his Macintosh partners, Henry Birley, had both made a point of testifying against Goodyear. Goodyear saw the victory as public vindication that he, rather than Hancock, was the sole inventor of vulcanized rubber. He lost no time in writing to

Stephen Moulton with the good news, taking the occasion to vent some pent-up resentment about Hancock's memoirs as well: "I am happy to say to you my patent is extended, enclosed. I send to you the testimony of Hugh Birley and Thomas Hancock. This will form a nice appendix to the in-famous Book of Hancock where the name Goodyear cannot be found." Goodyear continued: "Posterity will judge of the moral motives of the man when this evidence volunteered by the man himself is put together with the history of the facts in juxtaposition with that book and another written by Goodyear."

But fortune never smiled on Charles Goodyear without taking something away at the same time, subjecting him to some gross tragedy or small humiliation. The triumphant letter by a man who had set in motion a great industry and bested powerful foes in court, contained this postscript, scribbled hastily at the dockside: "I find that the evidence of Hancock and Birley will cost a dollar postage—will send the whole evidence in pamphlet by steamer another day."

15

LAST RITES AND WRONGS

In the winter of 1859, Goodyear made what would be his final move, to Washington, D.C. In sixty years the itinerant inventor had lived in a dizzying procession of homes in Connecticut, Pennsylvania, New York, Massachusetts, England, and France. Perhaps the family had found Washington a congenial place during Goodyear's patent extension fight. Or perhaps the Goodyears simply needed a change of scenery after enduring yet another tragedy. Year-old Arthur, born in Bath, died in New Haven in early 1859. Two years earlier, the couple's first child, Alfred, had died in Bath a month after his first birthday. By the time they moved to Washington, Fanny was already pregnant with a third. The family settled in a small house on I Street, a few blocks northwest of the White House. Though Goodyear's finances remained hopelessly tangled, Holt's decision had ensured that his patent would be protected and that he would continue to receive enough income to live on. His unmarried children by Clarissa, Clara, eighteen, and William Henry, twelve, lived with them. The boy, born in 1846, was the third to carry the name William, and the second William Henry.

Goodyear was able to take solace in Holt's decision and in his own renown. For years during his darkest struggles, he had been most terrified of the very real prospect that he would die before he could make his secret known—before he could prove to the world that rubber could be made impervious to heat and cold. He had seen his mission always as something more religious or spiritual than material. Goodyear did not fear death, but the thought of his great discovery languishing in obscurity was intolerable to him. But now vulcanized rubber belonged to the whole world.

Still, Goodyear remained restless, ever imagining new improvements and new products. To slake his incessant thirst for experimenting, he turned his attention to perfecting the life preserver. There was a marvelous symmetry in this choice for a final mission. After all, it had been a life preserver that ignited his interest in rubber more than a quarter of a century earlier. Now that his improvements in rubber had made durable, reliable life preservers possible, he set about trying to build one. In one room he installed a large water tank to test his devices.

On some level, though, Goodyear recognized that the long, great struggle of his life was nearing an end. He had persevered. And he had won. There were no more mountains to climb. A sort of internal peace came over him. Fanny Goodyear noted what she called "a marked ripening for glory" in Goodyear's disposition, "a growing gentleness and forbearance; an increased spirituality of mind, and a superiority to earthly care and anxiety."

In Washington, Goodyear was bedridden much of the time. Yet on one score his physical stamina always seemed more than adequate: on April 25, 1860, Fanny gave birth to a healthy girl, whom they named for her. Goodyear would have only five days to enjoy his new daughter before being greeted with distressing news. His daughter Cynthia, thirty-two, had fallen ill in New Haven. Cynthia had married a cousin, George Goodyear, in New Haven upon the family's return from England. Cynthia's condition, he learned, was dire. Despite his own poor health, Goodyear made hasty plans to leave for New Haven. Fanny would have to remain behind to care for the newborn. Goodyear's personal physician agreed to accompany the ailing inventor, but strongly advised against the rough overland journey. Instead, he left aboard the steamship *Montebello*, which steamed south down from the Potomac River and to the mouth of the Chesapeake Bay and then turned north to New York. Bad weather rocked the ship, and Goodyear added violent seasickness to his host of other ailments. He wrote a cheerful letter home to Fanny, saying, "Providence always smiles in the storm as well as in the sunshine." After a trip lasting several agonizing days, the *Montebello* docked in New York in early June.

George Goodyear, his son-in-law, met the ship with the news that Cynthia had died on May 31. Goodyear had fathered a dozen children by two wives, and now, with Cynthia's passing, had seen seven of them die. Cynthia was his second-oldest child after Ellen. Born during those fleeting days of happiness and prosperity in Philadelphia, Cynthia had survived a childhood of poverty and turmoil in one rubber-stained hovel after another. Now, settled in Connecticut, she had found a measure of happiness. Her death, four months shy of her thirty-third birthday, sent Goodyear into a spiral of grief too heavy to bear. Unable to continue his journey to New Haven for the funeral, he was taken to the Fifth Avenue Hotel. Installed in a bed at the hotel, he sent for his brother-in-law, William DeForest, in an effort to set his affairs in order, to the extent that this was possible. It must have been an emotional meeting between Goodyear and DeForest. Perhaps they reflected on a similar meeting many years before in Manhattan, when DeForest had visited Goodyear's walk-up apartment and chastised him severely for wasting his time on fruitless experiments while his family suffered. Of course, Goodyear had long since vindicated himself on that score—nobody could suggest that Goodyear's experiments had been wasted. And yet neither could the brothers-in-law exchange the satisfied smiles that might have come in the wake of unquestioned glory. For as with everything in Goodyear's life, the good news was attended always by bad; there were no absolutes, no complete vindication. The fact was that Goodyear's family not only had suffered greatly but continued to suffer. There would be no huge financial payoff for those who stuck with the inventor during all the difficult times. As DeForest prepared to leave, the doctor pulled him aside and confided, "This is the last."

When Fanny received news of Goodyear's condition in New York, she rushed to meet him despite having recently given birth. She arrived on June 7, finding the inventor even more feeble than she had expected. Sliding in and out of consciousness, Goodyear recognized Fanny and greeted her warmly, although he was unable to speak more than a few words at a time. He lingered on for the rest of the month, usually surrounded by family members. Appropriately for a man who believed himself to be on a mission from God, Goodyear died on a Sunday, July 1, 1860. He was six months shy of his sixtieth birthday.

Had Goodyear been able to arrange the circumstances of his own death, it's unlikely he would have changed much. His mission was complete; he had at last received widespread credit. He seems to have died with a remarkable lack of rancor. Fanny Goodyear would later report that the last words Goodyear uttered to her before dying was that he bore no ill will toward those who had stolen from him.

Despite his notoriety in his later years, Goodyear's death occasioned brief if respectful newspaper obituaries. The *New York Times* placed the cause of death as gout but, inexplicably, neglected to say anything about Goodyear's inventions or his life, or about any of the high-profile court cases that had occupied so much public attention. The *New York Tribune*, edited by Horace Greeley, was more generous in its treatment: "He felt the burden of honorable obligations which his patents imposed upon him, and, with complete disinterestedness, kept himself poor that he might enrich the world at the earliest possible moment with the greatest number and variety of completed practical applications of the material to man's uses." The obituary concluded that Goodyear had "been taken away in the midst of his useful labors."

The next Sunday, the Reverend Samuel Dutton delivered a lengthy sermon on Goodyear to a packed congregation at the North Church in New Haven. Dutton began by comparing the inventor with the biblical Bezaleel, son of Uri in Exodus, whom God fills with wisdom and understanding so that he might "devise curious works."

Dutton credited Goodyear with taking a product that was "almost worthless" and rendering it through long years of toil into "an article of inestimable value and of indispensable utility." Dutton said, "Already the various modes of mechanical industry founded upon it give employment to thousands and supply beneficently the wants of millions in all parts of the civilized world."

Dutton dealt forthrightly with Goodyear's atrocious financial habits: "He was, especially in his later years, improvident; so that, though in the receipt of large sums of money, he was yet often embarrassed with debt, to a degree which was a discomfort to his family and friends, and a disadvantage to his creditors." It was an honest assessment, necessitated perhaps by the fact that a good many of those family, friends, and creditors (often, they were one and the same) sat listening to Dutton's sermon. Like many contemporary critics, Dutton excused Goodyear's financial recklessness as the by-product of creative genius. "Still, his improvidence was a fault to be regretted," Dutton said. "His character would have been more complete if this had been otherwise."

In the end, Dutton found important lessons for others in the life story of Charles Goodyear: "In the first place, find out what your peculiar endowments are, what talents are entrusted to you, what you are called to do. Then, in the second place, do it—do it industriously and earnestly." Of Charles Goodyear no truer words could be said than that he found his life's calling and did it, industriously and earnestly.

Sitting in the congregation that day, listening to Dutton, were some of the people who had been most helpful to Goodyear in his times of need. One was Benjamin Silliman Jr., the Yale professor who had vouched for Goodyear's inventions from the earliest days of Goodyear's experiments. Goodyear was buried in a spacious plot in the northwest corner of Grove Street Cemetery, New Haven's most prominent and important burial ground. The raised sepulcher, bearing Goodyear's name in large letters, is a ponderous granite rectangle whose solidity suggests just the sort of robustness and prosperity that eluded him throughout his life. The man who hauled his ragged family from one town to the next, restlessly pursuing his dream, rarely stopping in one place for more than a year or two, is rooted for all time at this shady spot a block or two from the Yale campus, surrounded by the graves of former Yale presidents, distinguished scholars, decorated soldiers, and assorted dignitaries. Eli Whitney, the inventor of the cotton gin, lies a few hundred feet away from Goodyear, as does Benjamin Silliman Jr., who died in 1865.

Goodyear's affairs at his death were every bit as tangled as they had been during his lifetime. It fell to Goodyear's son to serve as executor. As Charles Jr. began to unravel the tangles of debt and income, the news was mostly bad for Goodyear's dependent survivors. The $33,358 in profits he claimed to have received since having been awarded the patent turned out to be a mirage, for he had simply overlooked mountains of debts. In the end, undisputed debts against the estate totaled $191,100.73. Among his creditors was his brother-in-law, William DeForest, whom he owed $30,000. William Ballard, an early supporter, claimed $300. The Connecticut Savings Bank wanted $4,500 to cover unpaid loans, while the Bank of Metropolis said Goodyear owed it $2,144.23. George Goodyear, his son-in-law, had lent the inventor $5,300 that was unpaid at the time of Charles's death. Goodyear owed G. P. A. Healy, the portrait painter, $670. And the list went on. Goodyear had made no provisions for paying off his debts, other than the willy-nilly, come-what-may approach that he had always used. He had no assets to speak of, save for the income from his patent licenses and the modest home in Washington. On the other side of the ledger were accounts receivable that Goodyear had never been able to collect, or never bothered to pursue. Chief among these was the account of Horace H. Day, whom Charles Jr. figured owed the inventor's estate nearly $8,000 in unpaid royalties for the shirred-goods license. Day denied that he owed the money, and never paid.

Dependents at Goodyear's death included Fanny and her infant daughter, as well as Clara and William Henry, his unmarried children by Clarissa. Goodyear ought to have been able to leave behind a fortune to keep his family in comfort, if not riches, for generations. Instead, there was a serious question about how his dependents would be able to survive. Charles Jr. proved to be an able administrator, managing to pay off most of the debts during the remaining life of the patent, and to put aside a modest but livable income for the dependents.

Thomas Hancock, though fourteen years older than Goodyear, lived for another five years, dying a wealthy man in London at the age of seventy-nine. He spent the last seven years of his life in comfortable retirement. He had much to be proud of: he could rightly claim to be one of the founders of a major industry; his ingenuity had contributed greatly to the understanding of rubber; tools he crafted were by now used universally; and his business skills had enabled him to turn a fledgling company into the most powerful rubber manufacturer in the world, creating jobs for thousands of workers over the decades. Upon his retirement from Macintosh & Company in 1858, grateful employees presented him with an elaborately printed testimonial. It said in part: "We the operators in the employ of Chas Macintosh and Company cannot permit the opportunity to pass of your retirement as partner in the above firm without expressing our heartfelt gratitude for the kindness, generosity, and benevolence which you have so liberally bestowed upon us while in your service." The testimonial thanked Hancock for his "Christian forbearance, mildness of counsel, and impartiality, which have assumed more the character of an indulgent parent than an employer." He had done much, his employees said, "to elevate the mind and raise the scale of morality."

They were sincere words, and heartwarming, no doubt, to their recipient. And yet the meditation about morality had to sting Hancock with unintended irony. Any mention of Hancock and morality by now came fixed with an asterisk, echoed in the measured tones of Lord Campbell's understated assessment in 1856. Hancock's conduct, when it mattered most, had been "not handsome."

Harper's Weekly published a lengthy article on Goodyear and Hancock while both men were still alive. The article studiously avoided taking sides in the vulcanization dispute, saying, "Of course, we do not pretend to adjudicate a dispute which has exercised so many lawyers' wits." But the underlying facts were not so difficult to assess, and even some men whom Hancock counted as friends took a dim view of his actions. Among the most damning assessments was one Hancock never saw, from Alexander Parkes, the eminent British chemist and a man Hancock particularly admired. Hancock had sent a copy of his memoirs to Parkes. Parkes, a gentle and gentlemanly soul, sent back a polite letter of congratulation. He could hardly do otherwise, since Hancock had singled him out for praise in the book. But in the margins of his personal copy, Parkes penciled in his private thoughts: "He must always have on his mind Goodyear, for to him alone is the invention due." In the section where Hancock details his vulcanization work, Parkes wrote: "This discovery was made first by Goodyear in America and not by Hancock. Justice must be done to Goodyear." In an ironic twist, Parkes would later find himself facing a similar situation in reverse, as American chemical manufacturers John and Isaiah Hyatt usurped credit for his invention of celluloid.

Parkes may have kept those thoughts to himself. But the sentiment was widespread on both sides of the Atlantic. Hancock could never bring himself to acknowledge Goodyear's discovery, nor mention his name in print. Even as he admitted that he had examined the samples from America, Hancock continued to insist that he was the inventor of vulcanization. "Now, reader, you may judge of the hardship of my case," he wrote pleadingly in his memoirs. "How could I *prove* that I was the first inventor? How could I *prove* that I made the invention at the time I applied for my patent? I could assert it, but how could I prove it?" Hancock's efforts to reconcile his inconsistencies grew ever more strained. He claimed that the American samples simply demonstrated to him that vulcanization was possible, but "they afforded no clue to the mode by which [vulcanization] had been brought about." He further claimed that he made no effort whatsoever to analyze the pieces. As for the presence of the sulfur bloom, Hancock said he thought sulfur had been rubbed onto the surface of the piece as a deliberate attempt to mislead the curious.

Here, then, was the essence of Hancock's explanation to the world. Since he got no clues from the samples, his decision to retreat to his private laboratory rather than credit the inventor was really nothing more than a sound and proper business decision, all in the spirit of a good race: "As it was my particular department to keep up the quality of our manufactures, and to maintain our standing and the position our goods had attained, I set to work in earnest, resolved, if possible, not to be outdone by any." It is difficult to imagine a man as intellectually curious as Hancock failing to analyze the most important rubber samples he had ever seen. Yet that is what we must believe if we are to believe Thomas Hancock.

If Hancock harmed his own reputation, his defenders compounded the problem following Hancock's death. It would have done his memory much better service for his defenders to recognize Goodyear as the inventor of vulcanization, quietly acknowledge events as they happened, and move on with due speed to celebrating Hancock's other achievements.

And yet as Goodyear's reputation swelled, a culture of defensiveness developed around Hancock, and a stubborn unwillingness among his defenders to acknowledge the ragged genius who beat Hancock to the draw, if not to the patent office; a need to make excuses for Hancock's inexcusable act. The most egregious example came in 1920, with the republication of Hancock's memoirs by James Lyne Hancock Ltd., named for a nephew who inherited Hancock's London factory. Vulcanization, the introduction states unequivocally, "was first given to the World by Thomas Hancock when he patented the process in 1843." The preface cites Goodyear for "experimenting in the same direction, and arriving at somewhat similar results, though by different methods." Most preposterously, it claims that Goodyear did not fully understand the process until after Hancock's patent.

Even in our own time, some argue that Hancock's actions represented sound competitive tactics in a competitive world, and that Goodyear was a fool to show his samples around. As rubber historian Austin Coates put it in 1987, "Goodyear had no one to blame but himself." While Goodyear certainly showed poor judgment, this line of reasoning is not far off from suggesting that a pickpocket is not a pickpocket if the victim ventured into a rough neighborhood with his wallet peeking out of his trousers.

Frank Duerden, a British rubber scientist, asked in a 1986 tribute to the Englishman, "If Hancock obtained some help from a knowledge of Goodyear's success then surely Goodyear himself owed a similar, although not identical, debt to Nathaniel Hayward?" But Goodyear always made a point of acknowledging his debt to Hayward. More important, Hayward, by his own admission, *did not* discover vulcanization. Sulfur was a vital step, but no more. Goodyear's persistence took care of the rest. Hancock, by contrast, reinvented Goodyear's invention. As for the frequent claim that Hancock's use of just sulfur and rubber (without Goodyear's white lead) represented a fundamentally new and different invention, Hancock's own actions contradict it. Hancock, after all, zealously (and successfully) fought imports of Goodyear-licensed shoes to England on the very basis that Goodyear's process was fundamentally the same.

Finally, Hancock defenders haul out the inevitable Goodyear stereotype—that of the fortunate bumbler. Even if Goodyear did stumble onto vulcanization, the reasoning goes, Hancock was the true scientist. "And was not Goodyear's discovery a chancy affair compared to Hancock's persistent and systematic experimentation?" Duerden asks. Coates compares Goodyear's "red-hot stove accident" with Hancock's "fundamental discovery reached by incalculable trial, observation, and experience." In the end, this argument does the greatest injustice to Goodyear, for it negates his years of dogged, persistent struggle both before and after the accidental discovery in Woburn. Thomas Hancock did not need an accident to point his way; he had Charles Goodyear. In the end, both men paid a terrific price for their actions. The cost to Goodyear was every worldly comfort; to Hancock, immortality. Whether, as Alexander Parkes had guessed, Hancock carried with him always thoughts of Charles Goodyear, only Hancock himself could know.

16

A WORLD MADE OF RUBBER

Charles Goodyear's U.S. patent extension lasted for five years after his death. As 1865 approached, the family, led by Charles Jr., decided to attempt one last extension. The patent had been under Goodyear's name for two decades, double the protection usually offered to patentees. With each passing year it became more and more difficult to prove that forces beyond the inventor's control had prevented him and his descendants from receiving their just rewards.

The list of licensees was long and impressive; it included the Union India Rubber Company, the Boston Belting Company, Henry Edwards and Partners, the Goodyear Elastic Fabrics Company, the New England Car-Spring Company, the Shoe Associates, Daniel Hodgmen, the New York Rubber Company, the Newark India Rubber Manufacturing Company, Goodyear India Rubber Glove Company, Nashuanneck Manufacturing Company, L. Candee (for "noiseless slate frames"), Nathaniel Hayward, John Haskins, the New York Belting and Packing Company, the New York Rubber Company, and the American Hard Rubber Company (for covering telegraph wire).

Charles Jr. emphasized the destitute relatives, "who are nearly all, if not all, wholly dependent on the settlement of his estate." He further argued his father's aims were always to further the understanding of rubber rather than to reap financial gain: "He carried on his experiments on a very extensive scale in divers places, almost on the scale of a manufacture, for which he never derived any reimbursement or remuneration." As to Goodyear's opus, *Gum-Elastic and Its Varieties*, the son pointed out that the book had never been formally published, that his father never considered the book to have been completed, and printed only a few copies "for his own private use and for the use of his friends. It has never been on sale."

Opposing the patent extension, Horace H. Day journeyed from New York to Washington for one last chance to harass Charles Goodyear in death as he had in life. By 1865, Day was all but out of the rubber industry, so his interest in the case could have amounted to little more than spite. But he could not contain his innate dislike for Goodyear. Joining Day in opposing the extension was an array of powerful interests whose very presence affirmed the crucial role that rubber was already playing in American industry. One of the loudest unified voices came from the railroads—the Pennsylvania Railroad; the Philadelphia, Wilmington and Baltimore; the Philadelphia and Reading; the Camden and Amboy; and a variety of others who in a few short years had come to rely on vulcanized rubber for belts, packing, car springs, and hose.

Day's attacks against Goodyear at the patent hearing were particularly harsh and spurious. Still clinging to his claim that Goodyear had not invented vulcanization, he accused the dead inventor of having perpetrated a "vast fraud upon the community," and he called him "the great, original confidence man." He claimed absurdly and irrelevantly that Hancock's English patent, since it was issued before Goodyear's American patent, rendered Goodyear's invalid. Without a shred of evidence beyond his own imagination, Day asserted that Goodyear had earned close to five million dollars from the invention, leaving erstwhile investors adrift in poverty while he enjoyed great luxuries: "He has enjoyed all the luxuries which wealth could give; he has in every conceivable way ministered to his vain glory, his appetites and his passions; and, as witnesses tell you, riding through the principal cities in Europe, in the style of nobles and kings; he has squandered money on the right and on the left, wholly indifferent to his moral or legal obligations."

Clarence A. Seward, an attorney for Charles Jr., blasted Day: "For twenty years and more did he unrelentingly pursue his benefactor. Ordinarily one buried animosity for an enemy in that enemy's grave. Mr. Day seems to be an exception. He has refused to let his wrath go down with the sun."

George Griscom, the lawyer for the opponents of the extension, must have drawn some twitters when he tried to defend Day: "Who can deny that Horace H. Day ranks equal to any man on earth as a thorough-bred, correct and irreproachable merchant, and as an energetic and successful manufacturer?"

The hearings provided some additional comic relief when Day took it upon himself to interrogate Charles Jr. on the witness stand. The son's barely concealed contempt shows through in an exchange that reads like an Abbott and Costello routine:

DAY: Name any years in which you worked in India rubber factories, and where.

GOODYEAR: I have worked in India rubber factories in Naugatuck, New Haven, New Brunswick, Newark, Hamden, CT; and also at Providence. I decline to answer as to dates.

DAY: Why do you decline to answer as to dates?

GOODYEAR: I decline to answer.

DAY: When was the first factory erected, to your knowledge, in Naugatuck? State the time as nearly as you can.

GOODYEAR: I decline to answer.

DAY: Do you know?

GOODYEAR: I decline to answer.

DAY: Did you ever reside with your father's family at Springfield, in Massachusetts?

GOODYEAR: I did.

DAY: When?

GOODYEAR: I decline to answer.

DAY: Why do you decline to answer this question when you must see its direct bearing upon the question of the knowledge necessary to enable you to fairly present your father's case before Congress?

GOODYEAR: I decline to answer.

The exchange must have warmed the hearts of the Goodyear contingent—the inventor's son, with Bartleby-like indifference, rebuffing questions put to him by the arrogant, white-maned, and increasingly red-faced Day. In the end, though, the cause of the heirs was doomed, not by Day's machinations, nor even the powerful forces against the extension, but by the patent office's reluctance to extend a patent that had already been held for twenty years. The patent office refused the request, announcing that Charles Goodyear's patent for vulcanized rubber would expire on June 15, 1865. The invention that he had always seen as his gift (or, rather, God's through him) would now be, in the truest sense, just that. Vulcanized rubber belonged to everybody.

Horace Day outlived his principal enemy by eighteen years. During the course of his long and combative career, Day had made many enemies in addition to Charles Goodyear. Those who longed to see Day finally get his comeuppance at last got their wish. It took the Niagara Falls to bring him down. After his expulsion from the rubber industry, Day became obsessed with harnessing the energy generated by the falls. He spent about $700,000 for seven hundred acres of land adjacent to them. The previous owner of the property had started but not finished a large canal to power factories. Day established the Niagara Falls Water Power Company, installed himself as vice president, treasurer, and director, and finished the canal.

The project was hugely expensive, forcing Day to charge extremely high rent for the land. If he had imagined himself the baron of a vast manufacturing fiefdom, the reality was that he had few or no takers at the rates he was forced to set. Day's combative personality made him particularly ill-suited to negotiating with prospective tenants. When one manufacturer offered $20,000 a year in rent for a parcel, Day, instead of negotiating, angrily rejected the offer out of hand. Day found himself the lord of an empty and costly kingdom. Without tenants, just meeting the debt service on his land rapidly shrank his fortune. In desperation, he began to devise ever wilder schemes for the land. One plan called for harnessing compressed air from the falls in a pipe that would deliver six thousand horsepower for factories in Buffalo, twenty miles away, or even in New York City. Although engineers said the plan was theoretically feasible, Day couldn't find the financial backing. Next, he tried to stir up interest in a "marine railway" that would carry ships around the falls on giant railroad cars. He rounded up favorable opinions from prominent engineers in a pamphlet he published in 1865, stating that the marine railway was a grand idea. Once again, investors shrugged. Day was forced to take out repeated mortgages on the property, sinking him further into debt. As his financial condition worsened, Day sought solace ever more fervently in the spirit world. There was one spirit he wished to contact above all others—that of his erstwhile courtroom foe Daniel Webster. Webster's spirit, speaking through the voice of an eager-to-please medium, assured Day that he would become solvent again.

While Day communed with ghosts, his fortunes in the material world sank. In the end, he was forced to sell his Niagara property at auction for just over $70,000, about a tenth of what he had paid. The buyer agreed to assume Day's enormous debts. He slunk back to New York from Niagara Falls a comparatively poor man.

Day's next venture put an appropriate coda on his notorious career. By 1874, the Porous Plaster Company of New York had been selling its Allcock's Porous Plaster curatives from its Canal Street factory for nine years. The plasters, applied directly to the skin for as long as three months, could, according to its manufacturers, cure any number of maladies, by restoring balance to the body's electrical system. Consumption? Apply a patch over the breastbone. Kidney pain? The lower back, on the side. Diarrhea or bowel problems? Place one "over the bowels." One could cure ague, spinal troubles, blisters, asthma, liver disease, and menstrual cramps.

Proprietors of the Porous Plaster Company were considerably irked, in 1874, to see plasters in suspiciously similar wrappings and of suspiciously similar type appearing on the market. The bogus plasters had circular labels with an eagle and clouds at the top, just like theirs, along with the same patent date and instructions for wetting the muslin to remove the plasters. Printed on the packages was the following message: "The undersigned is the original Inventor and Patentee of the INDIA RUBBER POROUS PLASTER, and has resumed the manufacture . . ."

The name below the message was Horace H. Day.

The company sued. In May 1875, a New York State judge found for the plaintiffs. Day was ordered to quit making porous plasters and was fined fifty dollars. The case, and even the meager nature of the fine, underscored a rather pathetic end for a man whose infringements had once drawn the attention of the greatest legal minds of the age. Financially broken, his athletic physique beginning to fail him, Day left New York for Manchester, New Hampshire. He was working on a method for grinding tree bark to be used in tanning leather when he died on August 24, 1878. He was sixty-five. The New Hampshire death record listed the cause as inflammation of the bowels—one of the conditions that his porous plasters were supposed to correct. The *New York Tribune* described Day as a man "once widely known as a manufacturer and promoter of speculative enterprises," and concluded with this assessment: "In his intercourse with men he was very dictatorial and arrogant."

In 1861, the year after Goodyear died, Union armies marched off to battle wearing rubber ponchos and pitching rubber tents to protect themselves from the elements. Battlefield doctors carried rubber medicine bottles and soldiers carried rubber powder flasks. By the late 1800s, a material once forgotten was vital enough to drive mild-mannered dentists to murder. On Easter Sunday, 1879, Josiah Bacon, treasurer of the Goodyear Dental Vulcanite Company, was found dead in his San Francisco hotel room. The company—unrelated to the Goodyear family—owned the patent for hard rubber dentures, which had revolutionized the dental industry by replacing ill-fitting wooden dentures with snug rubber. Prompted by Bacon's bulldog tactics, the Goodyear Dental Vulcanite Company exacted usurious fees from dentists across the country and zealously tracked down and prosecuted any and all who violated the patent. Dentists claimed they had no choice but to violate it—patients demanded vulcanite, but dentists could not afford the steep fees. In a fit of anger Dr. Samuel P. Chalfant, a defendant in one of Bacon's lawsuits, tracked the treasurer down and assassinated him. Chalfant served ten years in prison before being released. He continued his practice, and to this day remains something of a hero in the dental community.

Vulcanization created a worldwide demand for South American rubber. What had once been a sleepy trade of a material in little demand now created vast fortunes for some and torment and abuse for thousands of others. Grandiose opera houses sprouted in Brazilian jungle towns such as Manaus, along with saloons where the finest French champagne could be bought at four in the morning. The Brazilian rubber boom ended almost as suddenly as it began, when an English adventurer named Henry A. Wickham spirited away seventy thousand *Hevea brasiliensis* seeds from Brazil. The seeds were planted in Kew Gardens, outside of London. Some two thousand seedlings were then transported to the Far East. The move created the plantation rubber industry in Asia and in one stroke effectively killed the Brazilian rubber industry.

By the dawn of World War II, rubber was so important that Japanese control of Far Eastern plantations threatened the Allied war effort. Rubber was needed for vehicle and airplane tires, rubber rafts, and rubber lining for gas tanks, to seal bullet holes. It was used for boots and boats and pontoon bridges. The United States led a massive effort to create a usable synthetic rubber. By the end of the war, synthetic rubber was well on its way to overshadowing natural rubber. And yet every ounce of synthetic rubber still had to be vulcanized in much the same way as pioneered by Charles Goodyear.

Goodyear's name lived on in the rubber industry because of the many licensees that adopted it. A casual observer would have assumed a family dynasty in the making, but Goodyear's descendants had little to do with any of it, and received little financial reward. Charles Jr. had assisted his father in life and was charged with the unenviable task of straightening out his estate after his death. But he dropped out of the rubber business not long after his father died. He would make his mark as an inventor in the shoe industry. He developed the Goodyear Welt, a mechanized method for fixing the upper shoe to the sole that is still widely used today. The shoes I am wearing as I write say "100% Caoutchouc" (the old name for rubber) on the outer sole and "Goodyear Welt" on the insole, a moving if unintended tribute to the combined talents of father and son.

Another son, William Henry, grew up to became a renowned art historian, author of several influential books, and curator at the Metropolitan Museum of Art in New York and the Brooklyn Institute Museum. The Goodyear invention streak would take a final, odd twist with granddaughter Anna Goodyear, a severe-looking woman who ran a mission in Boston's North End. In 1907 Anna devised a spanking machine for adult criminals. The rotating platform could be adjusted to the height of the offender, and set for one to fifty blows. Anna insisted the humiliation of a public spanking would be more effective than imprisonment in rehabilitating offenders. She offered to present a free machine to any town or city that wanted one, but there were few takers, even at such a reasonable price.

The force that did more than any other to advance Goodyear's name and cement it to his invention in the public mind had no formal connection with the man or his family. Frank Seiberling of Akron, Ohio, born a year before Goodyear's death, had spent twenty years in various enterprises, including a twine and cordage factory, mills, and a streetcar company. Nearly wiped out in a national financial panic started by the failure of the Philadelphia and Reading Railroad, Seiberling by 1898 was looking for a new opportunity. He scraped together the down payment for a defunct Akron factory on twelve acres without knowing what exactly he would manufacture.

Bicycles were hugely popular, and a new horseless carriage seemed promising. Akron was already home to two established rubber manufacturers, Diamond Rubber Company and B. F. Goodrich. Perhaps the growing city could support one more. Seiberling's brother, Charles, joined him in the venture. Instead of naming the company after themselves, the brothers selected Goodyear's name, to honor, they said, the inventor's perseverance and dedication. The name also had an agreeable similarity to the more established Goodrich company. The Goodyear Tire & Rubber Company opened its doors November 21, 1898, with thirteen employees.

As Goodyear Tire & Rubber grew into the world's largest tire company, a true manufacturing colossus, the relationship between the company and its namesake remained a peculiar one. Many companies have dealt with a legacy of eccentric founders. But few have been so closely associated with a man who was *not* the founder. To this day most people assume a direct connection between Charles Goodyear and the tire company, an assumption that generally works to the favor of both.

Because of its name, the company over the years has become the heir to Goodyear lore and memorabilia. The somewhat threadbare World of Rubber exhibit, located on the fourth floor of Goodyear Hall in Akron, features a mockup of a shack where a life-size Charles Goodyear toils at his invention over a potbellied stove. Nearby are glass cases exhibiting various medals awarded Goodyear, some manuscripts, buttons, combs, and other objects the inventor fashioned out of hard rubber, portraits on hard rubber by Healy, and other items.

A bronze statue of Charles Goodyear, unveiled during the company's massive celebration of the hundredth anniversary of vulcanization, still gazes over Akron from its leafy perch in a downtown park. In a rather bizarre bit of self-congratulation during a company dinner in 1925, top executives watched Goodyear return from the grave (as projected through a company flack). He told them: "I have come back tonight to visit the great structure that you have erected in my name. I have come back in pride—because you have not belied your heritage, because you have pursued the goal of service to mankind, because you have been unafraid, because you have the faith to move mountains."

Over the years, the company has walked a fine line between embracing the Goodyear legacy and making it clear that there is no *legal* connection between the firm and the inventor or his family. Indeed, none of Goodyear's descendants have ever had a prominent position in the company. The company was given to acts of generosity with some descendants. When grandson Charles Goodyear III and granddaughter

Clara Goodyear (married cousins) fell on hard times during the 1930s, the company quietly arranged for a monthly stipend of one hundred dollars per month, later doubled, for the rest of their lives. At other times, the company strove to keep relations at arm's length. Granddaughter Rosalie Eliot Heaton, irked when the company refused to buy a pair of rubber cuff buttons worn by Goodyear at the court of Napoléon III, began writing angry public letters to company directors. "Grandfather's life was faithful, just and righteous," she wrote. "But our family has had no evidence of concrete tribute being paid to us either for Goodyear's name or process which has been used for some sixty years."

Whatever Goodyear descendants have thought of the connection, it is difficult to imagine Charles himself being anything but pleased. He would have been gratified to see his name on millions of vehicle tires. Especially, though, the aesthete in him would have loved the blimp that carries the Goodyear name to sporting events around the country. A man who always saw his quest as a holy mission would have adored seeing his name hovering godlike over great stadiums, where thousands of rubber-soled spectators gather to watch grown men and women play games with balls made of vulcanized rubber.

ACKNOWLEDGMENTS AND SOURCES

This project began when I happened across Charles Goodyear's name in Webster's American Biographies *and assumed, like everyone else, that he had started the tire company. My journey from ignorant curiosity was a long and personally rewarding one, and I owe thanks to people and institutions on both sides of the Atlantic.*

Joan C. Long, the librarian for the Rubber Division Library in Akron, Ohio, set me on my way. She helped me find crucial books and documents, and guided me to scientists, historians, and materials around the country. John Miller, University of Akron archivist, offered me workspace over several days to examine two large boxes of Goodyear documents, clippings, and letters that provided much important information. The Goodyear Tire & Rubber Company, despite its lack of a formal connection with the inventor, has long been a de facto repository for Charles Goodyear collectibles. Fred D. Haymond and Donna L. Jennings, from the public relations department, arranged for me to examine Charles Goodyear's London address book, Fanny Goodyear's medical diary, and other original items. Howard L. Stephens, emeritus professor of polymer science at the University of Akron, provided his expertise on the chemical properties that make rubber unique. Benjamin Kastein, a dean of American rubber historians, braved an Akron snowstorm to meet me for coffee and a spirited discussion of rubber at a restaurant near his home.

Staff members of the famed Science Museum in London went out of their way to help me. Mandy Taylor of the museum's library led me to a basement archives room where we retrieved a box containing hundreds of ancient glass negatives—undisturbed for years—featuring letters, portraits, and documents pertaining to Thomas Hancock and Charles Macintosh & Company. These were crucial in helping me round out Hancock's side of the story. Dr. Susan Mossman, curator of materials science, generously devoted an entire afternoon from her busy schedule to take me across London to the museum's main storage area, where she showed me Hancock's prototype masticator, samples of early rubber products, and other artifacts, and offered her expert commentary on each. Dr. Peter Morris, senior curator, shared his knowledge and offered useful tips for further research.

In the storybook town of Bradford-on-Avon, in Wiltshire County, England, Dr. Alex Moulton, great-grandson of Stephen Moulton, provided an afternoon of tea and conversation about his family history, on the terrace of the Hall, his ancestral home. At the Wiltshire & Swindon Record Office in nearby Trowbridge, where the Moulton Company records are now stored, the friendly staff deposited at my table a treasure trove of handwritten letters by Charles Goodyear, Charles Jr., Stephen Moulton, and others, as well as Moulton company records, depositions from legal battles with Hancock, and other invaluable information. Though I met her only by phone, Kristina Lawson of the Tun Abdul Razak Research Centre in Hertford, England, kindly tracked down and mailed me very helpful journal articles on Charles Macintosh and Michael Faraday.

I would like to thank Valerie Jablonowski, MaryLou Burmeister, Dana J. Blackwell, Verna A. Blackwell, Mary Doback, Ann Simons, and Helen Wilmot of the Naugatuck Historical Society in Naugatuck, Connecticut, for sharing what they know about Goodyear and the early Naugatuck rubber industry. Valerie Jablonowski gave me a comprehensive guided tour of the town, and I watched her husband, rubber chemist Tom Jablonowski, re-create Goodyear's vulcanization discovery.

Woburn, Massachusetts, historians Tom Smith and John D. McElhiney devoted a Saturday to tromping around East Woburn with me, bringing the nineteenth-century village back to life before my eyes, and even buying me lunch. Tom Smith generously shared research materials for his book, The India Rubber Man in Woburn. *McElhiney's* Woburn: A Past Observed, *must set a standard for thorough and elegant town histories. Thanks also to Roger D. Joslyn, a professional genealogist, who took on Horace Day as a personal mission and turned up some wonderful information.*

Thanks to several libraries and their staffs, who opened their doors and collections to me and were unfailingly generous during my visits. In particular: Rutgers University's Special Collections room, home to thick volumes of testimony relating to the Great India Rubber Case of 1852; Harvard University Libraries; The Library of Congress; the New Haven Colony Historical Society; the Springfield (Mass.) Historical Society; James Branch Cabell Library at Virginia Commonwealth University; Old Dominion University Library; the Virginia Historical Society; and the American Antiquarian Society.

All of these people and institutions shared their knowledge generously. I alone am responsible for any errors of judgment or fact.

Among the books listed in my bibliography, a few deserve particular mention. Of several Goodyear biographies that appeared around 1939—the centennial of his great discovery—P. W. Barker's privately published Charles Goodyear: Connecticut Yankee and Rubber Pioneer, *was especially succinct, factual, and unsentimental. It provided a helpful and reliable guidepost during my research and writing. William Woodruff's* The Rise of the British Rubber Industry During the Nineteenth Century *(1958) was clear and informative and first alerted me to the existence of the Stephen Moulton archives. And of course, there are the memoirs by Goodyear and Hancock, which, for all their omissions and stylistic quirks, offered indispensable windows on the characters of these two very different men. I would like to thank my agent, Andrew Blauner, for his wise guidance and hard work on behalf of this project; my editor, Gretchen Young, for her enthusiasm, careful reading, and important suggestions; her able assistant, Natalie Kaire; and the entire staff at Hyperion for showing such interest in this project.*

Finally, I would like to thank several people. Foremost is my wife, Barbara, who read a chunk of the manuscript by candle when a storm knocked out the power. Thanks to my daughters, Natalie and Caroline, and to my parents, Warner and Carolyn Slack, for support in so many ways. I must also thank Dean King, friend and mentor; Robert Hancock, Dana Goodyear, David A. Chernin, Claudio Phillips, Joann DiPanni, and Anthony DiPanni, for help and support in various ways.

BIBLIOGRAPHY

Books

Allen, Hugh, *Charles Goodyear: An Intimate Biographical Sketch,* Goodyear Tire & Rubber Co., Inc., Akron, Ohio, 1939.

Babcock, Glenn D., *History of the United States Rubber Company: A Case Study in Corporation Management*, Indiana University Graduate School of Business, 1966.

Barker, P. W., *Charles Goodyear: Connecticut Yankee and Rubber Pioneer,* Godfrey L. Cabot Inc., Boston, 1940.

Blackwell, Dana J., *Naugatuck,* Arcadia Publishing, Dover, New Hampshire, 1996.

Bryson, Bill, *Made in America: An Informal History of the English Lanugage in the United States*, William Morrow and Co., New York, 1994.

Chandler, Dean, and A. Douglas Lacey, *The Rise of the Gas Industry in Britain*, British Gas Council, 1949.

Coates, Austin, *The Commerce in Rubber: The First 250 Years,* Oxford University Press, Singapore, 1987.

De Mare, Marie, *G. P. A. Healy, American Artist*, 1934. Excerpt contained in the University of Akron Archives.

Dutton, Rev. Samuel W. S., *A Discourse, Commemorative of the Life of Charles Goodyear, The Inventor*, Thomas J. Stafford, New Haven, Conn., 1860.

Essays of Howard: Or, Tales of the Prison, The Literary Exchange, New York, 1811. Originally published in the New York *Columbian* newspaper.

Franklin, Fabian, *The Life of Daniel Coit Gilman*, Dodd, Mead & Co., New York, 1910.

Frémont, John Charles, *Memoirs of My Life,* Belford, Clarke & Co., Chicago, 1886.

Geer, William C., *The Reign of Rubber*, The Century Co., New York, 1922.

German, Reg, *The World of Rubber*, the British Rubber Manufacturers' Association, 1992.

Goodyear, Charles, *Gum-Elastic and Its Varieties*, privately published, New Haven, Conn., 1853, 1855. Reprinted by the American Chemical Society, 1939.

Goodyear, George F., *Goodyear Family History (Part III)*, privately published, Buffalo, N.Y., 1976.

Green, Constance McL., *History of Naugatuck, Connecticut*, privately published, Naugatuck, Conn., 1948.

Green, Mason A., *Springfield 1636–1886: History of Town and City*, C. A. Nichols & Co., Springfield, Mass. 1888.

Hancock, Thomas, *Personal Narrative of the Origin and Progress of the Caoutchouc or India-Rubber Manufacture in England*, Longman, Brown, Green, Longmans & Roberts, London, 1857. Reprinted by the American Chemical Society, 1939.

———, 1920 edition of original, published by James Lyne Hancock Ltd., London.

Healy, George P. A., *Reminiscences of a Portrait Painter*, A. C. McClurg and Co., Chicago, 1894. Reprinted by Kennedy Graphics Inc., Da Capo Press, New York, 1970.

Huke, D. W., *Introduction to Natural and Synthetic Rubbers*, Chemical Publishing Co., Inc., New York, 1961.

Illustrated American Biography, vol. III, J. Milton Emerson & Co., New York, 1853–1855.

In and Around Bath, Pitkin Unichrome Ltd., Andover, England, 1998.

Kirkman, Grace Goodyear, *Genealogy of the Goodyear Family*, privately published, San Fransisco, 1899.

Lewis, Walker (ed.), *Speak for Yourself, Daniel: A Life of Webster in His Own Words*, Houghton-Mifflin Co., Boston, 1969.

Long, Harry, *Basic Compounding and Processing of Rubber*, Rubber Division, American Chemical Society, Akron, Ohio, 1992.

McElhiney, John D., *Woburn: A Past Observed*, Sonrel Press, Woburn, Mass., 1999.

McLaughlin, James A., *Charles Goodyear of Woburn, Massachusetts* (booklet), 1932.

Morris, Peter J. T., *The American Synthetic Rubber Research Program*, University of Pennsylvania Press, Philadelphia, 1989.

———, *Polymer Pioneers: A Popular History of the Science and Technology of Large Molecules*, Beckman Center for the History of Chemistry, Philadelphia, 1986.

Morton, Maurice, *Rubber Technology*, 3rd ed., Chapman & Hall, London, 1995.

Mossman, Susan (ed.), *Early Plastics: Perspectives, 1850–1950*, Leicester University Press, London, 1997.

Nichols, Thomas Low, M.D., *Forty Years of American Life: 1821–1861*, Stackpole Sons Publishers, 1874. Republished 1937 by The Telegraph Press.

Parker, Edward G., *Reminiscences of Rufus Choate,* Mason Brothers, New York, 1860.

Peirce, Rev. Bradford K., *Trials of an Inventor: Life and Discoveries of Charles Goodyear*, Carlton & Porter, New York, 1866.

Penal Code of the State of Georgia, as Enacted December 19, 1816, With Reflections on the Same, and On Imprisonment for Debt., M. Carey & Son, Philadelphia, 1817.

Perry, Josephine, *The Rubber Industry*, Longmans, Green & Co., New York, 1941.

Remini, Robert, *Daniel Webster: The Man and His Time*, W. W. Norton & Co., New York, 1997.

Report on the Debtor's Apartment of the Arch Street Prison in the City of Philadelphia, Henry Welsh, printer, Harrisburg, Pa., 1833.

Rodengen, Jeffrey L., *The Legend of Goodyear: The First Hundred Years* (History of the Goodyear Tire & Rubber Co.), Write Stuff Syndicate Inc., Fort Lauderdale, Fla., 1997.

Roe, Joseph Wickham, *Connecticut Inventors*, Yale University Press, New Haven, Conn., 1934.

Russell, W. F., *The Vanderbilt Rubber Handbook*, 7th ed., R. T. Vanderbilt Co., New York, 1936.

Schidrowitz, P., and T. R. Dawson (eds.), *History of the Rubber Industry*, W. Heffer & Sons Ltd., Cambridge, England, 1952.

Sellers, Charles, Henry May, and Neil R. McMillen, *A Synopsis of American History*, volume I, *Through Reconstruction*, 4th ed., Rand McNally College Publishing, Chicago, 1976.

Shumway, Floyd, and Richard Hegel, *New Haven: An Illustrated History*, The New Haven Colony Historical Society, Windsor Press, 1981.

Smith, Tom, *Goodyear: The India Rubber Man in Woburn*, Black Flag Press, Woburn, Mass., 1986.

Stevens, Henry P., and W. H. Stevens, *Rubber Latex*, the Rubber Growers' Association, London, 1936.

Tocqueville, Alexis de, *Democracy in America, in Two Volumes with a Critical Appraisal of Each Volume by John Stuart Mill*, reprint by Shocken Books, New York, 1961.

Wall, John P., *The Chronicles of New Brunswick, N.J., 1667–1931*, Thatcher-Anderson Co., New Brunswick, N.J., 1931.

Wolf, Ralph F., *India Rubber Man: The Story of Charles Goodyear*, The Caxton Printers, Ltd., Caldwell, Idaho, 1939.

Woodruff, William, *The Rise of the British Rubber Industry During the Nineteenth Century*, Liverpool University Press, Liverpool, 1958.

Wyckoff, William C., *American Silk Manufacture*, the Silk Association of America, New York, 1887.

Newspapers, Journals, Pamphlets, and Other Documents

"Application of Nathaniel Hayward for an Extension of His Patent

Before the Committee on Patents," Washington, Gideon & Pearson, Printers, 1865.

Aurora & Pennsylvania Gazette (Philadelphia), Wednesday, July 1, 1829.

Boston Sunday Post, Sunday, September 1, 1907.

"Charles Goodyear: Phrenological Characters and Biography," *American Phrenological Journal*, December 1856.

Circuit Court, In Equity, Charles Goodyear Against Horace Day, depositions. From the collections of Rutgers University.

The Connecticut Journal, December 25, 1800, January 24, 1805.

Duerden, Frank, "Thomas Hancock: An Appreciation," *Plastics and Rubber International*, June, 1986.

Gilman, Daniel Coit, letters to Charles Goodyear, February 1854 to September 1855, from the Connecticut State Library collection.

Johnson, Roy, "House Where Goodyear Devised Vulcanizing in Danger of Demise," *Boston Sunday Globe*, September 7, 1969.

Meyer, L. Otto P., "Some Memories of Goodyear," *The India Rubber World*, August 1, 1901.

O'Reilley, Brian, "1852 Was a Goodyear in Trenton," *Daily Trentonian*, December 5, 1976.

Parton, James, "Charles Goodyear," *North American Review*, July 1865.

Philadelphia Democrat Press, January 20, 1827.

"Restoration Attempts by Many Fail; Charles Goodyear House Demolished," *Rubber & Plastics News*, April 24, 1972.

Ring, Malvin E., "The Rubber Denture Murder Case: The True Story of the Vulcanite Litigations," *Bulletin of the History of Dentistry*, April 1984.

Rodriguez, Tim D., "The Commercialization of Natural Rubber—A Historical Overview," paper presented at a meeting of the Rubber Division, American Chemical Society, Chicago, April, 1999.

Schurer, H., "Michael Faraday and Early Rubber Science," *Rubber Journal*, May 1967.

Silk: Its Origins, Culture, and Manufacture, Corticelli Silk Mills, Florence, Mass., 1911.

"The Story of India Rubber," *Harper's Weekly*, April 4, 1857.

White, James Lindsay, "Charles Macintosh of Glasgow: His Enterprises and the Foundation of the Rubber Industry," *Rubber Industry*, August 1974.

"World-wide Goodyear Joins in Home Coming Celebration," *The Goodyear News* (company newsletter), March 1939.

Woshner, Mike, "Focus: Civil War Plastic," *North South Trader*, November–December 1973.

INDEX